Raising Chicken for Beginners

From Coop Design to Egg Collecting

Emma Bennett

Table of Contents

INTRODUCTION

Welcome to "Raising Chickens for Beginners: From Coop Design to Egg Collecting." This book is a guide and a gateway to a new, rewarding lifestyle. Whether you're after the satisfaction of fresh eggs, the sustainability of a backyard farm, or the simple joy of having chickens, this comprehensive guide is the key to starting your chicken-raising journey. It's designed with newcomers in mind, demystifying every aspect of chicken care to make it accessible and enjoyable for you.

This book begins by helping you choose the perfect chicken breeds for your needs, considering factors like egg production, temperament, and climate adaptability. It then guides you through the essentials of designing and building a coop that ensures your chickens' safety and comfort. You'll create a haven for your flock with clear, step-by-step instructions.

Nutrition and feeding regimens are explained in detail, ensuring your chickens receive balanced diets for optimal health. You'll learn about common health issues, preventive measures, and basic first aid, equipping you to keep your flock thriving. Understanding chicken behavior and social dynamics will help foster a harmonious and happy flock.

As your chickens start laying, you'll master the best egg collection and storage practices. Seasonal care tips will prepare you for year-round chicken keeping, and advanced topics like breeding and raising chicks will enable you to expand your flock confidently. By the end of this book, you'll be well-prepared to enjoy the rewards of raising chickens, from fresh eggs to the simple pleasure of watching your chickens' roam. Welcome to the fulfilling and rewarding world of chicken raising!

CHAPTER I

Getting Started

The Joy and Benefits of Keeping Chickens

Keeping chickens is an increasingly popular pastime that offers a multitude of joys and benefits. For many, raising chickens conjures images of a more straightforward, more connected way of life, where the rhythms of nature are more closely observed and appreciated. The joys of keeping chickens are multifaceted, encompassing the practical, the emotional, and the environmental. This section explores these dimensions, highlighting why more people adopt this rewarding practice.

One of the primary benefits of keeping chickens is the supply of fresh eggs. Home-raised eggs are superior in taste and nutritional value compared to store-bought varieties. Chickens can roam freely and forage to produce eggs with richer yolks and more robust flavors. The nutritional benefits are also significant; free-range eggs typically contain higher levels of vitamins A, D, and E and omega-3 fatty acids. For families concerned about food quality and health, raising chickens provides a direct source of high-quality protein and other essential nutrients.

Beyond the practical benefits of egg production, chickens play a vital role in pest control. Chickens are natural foragers, and their diet includes a wide variety of insects, including ticks, beetles, and other pests that can be problematic in gardens and yards. By keeping a flock of chickens, homeowners can reduce their reliance on chemical pesticides, leading to a healthier and more sustainable living environment. Chickens also help with weed control, as they scratch and peck at the ground, consuming weed seeds and thus preventing them from taking root.

The environmental benefits of keeping chickens extend to waste management and soil health. Chickens produce a significant amount of manure, an excellent fertilizer source for gardens. Chicken manure is rich in nitrogen, phosphorus, and potassium—essential nutrients that plants need to thrive. When adequately composted, chicken manure can significantly enhance soil fertility, leading to more productive vegetable gardens and healthier landscapes. This natural recycling of nutrients exemplifies sustainable living practices and reduces the need for synthetic fertilizers, which can have harmful environmental effects.

In addition to the tangible benefits, keeping chickens provides significant emotional and psychological rewards. Many people find that caring for chickens brings a sense of peace and fulfillment. The daily routine of feeding and tending to the flock can be a grounding experience, offering a welcome respite from the fast-paced and often stressful modern life. Chickens are also known for their quirky and endearing behaviors. Each bird has its personality, and their antics can be a source of entertainment and joy. Watching chickens roam, interact, and express their natural behaviors is enjoyable and educational, especially for children.

Keeping chickens can foster a greater sense of community and connectedness. Sharing surplus eggs with neighbors or involving family and friends in the care of the flock can strengthen social bonds. In many communities, chicken keeping has led to forming local groups and co-ops where enthusiasts can share tips, resources, and support. These interactions can build a sense of belonging and shared purpose, enhancing overall well-being and community resilience.

For families with children, raising chickens offers unique educational opportunities. Through hands-on experience with the birds, children can learn about biology, ecology, and responsibility. They can observe the life cycle of chickens, from egg to chick to adult, and gain a deeper understanding of where their food comes from. This practical education can instill values of stewardship, empathy, and a love for nature that can last a lifetime.

Another significant joy of keeping chickens is its connection with nature and the outdoors. Raising chickens encourages more time spent outside in a world where many people spend most of their time indoors and disconnected from the natural world. The daily care routines—feeding, watering, cleaning, and collecting eggs —require regular interaction with the outdoor environment. This increased exposure to fresh air, sunlight, and natural elements can have numerous health benefits, including reduced stress and improved mental clarity.

Furthermore, keeping chickens can also be a form of self-sufficiency and empowerment. In an era where food security and sustainability are growing concerns, raising chickens offers a way to produce food. It can reduce dependence on commercial food systems, which are often vulnerable to disruptions. Providing fresh, nutritious eggs and managing a small flock gives individuals a sense of control and accomplishment, contributing to a more sustainable and self-reliant lifestyle.

While the benefits of keeping chickens are substantial, it is essential to recognize that this practice requires commitment and responsibility. Chickens need proper housing, regular feeding, and protection from predators. Potential chicken keepers must be prepared for these demands and willing to invest time and effort into caring for their flock. However, the rewards are plentiful and multifaceted for those willing to take on these responsibilities.

In conclusion, the joys and benefits of keeping chickens are diverse and significant. From the practical advantages of fresh eggs, pest control, and natural fertilizer to the emotional and psychological rewards of connection, community, and education, raising chickens enriches lives in numerous ways. It fosters a closer relationship with nature, enhances self-sufficiency, and promotes sustainable living practices. Keeping chickens offers a fulfilling and joyful path for those looking to improve their quality of life, connect with nature, and contribute to a more sustainable future.

Understanding the Commitment and Responsibilities

Understanding the commitment and responsibilities associated with any endeavor is paramount to its success and sustainability, and this is especially true when it comes to caring for living beings such as animals. Whether embarking on the journey of pet ownership, starting a family, or taking on a new job, recognizing the dedication and accountability required is essential. When it comes to commitments involving animals, such as keeping pets or raising livestock, the stakes are even higher, as the well-being and welfare of these creatures depend entirely on the actions and decisions of their caretakers. Therefore, it is crucial for individuals to thoroughly understand the commitment and responsibilities involved before taking on such roles.

First and foremost, the commitment to caring for animals encompasses their basic needs for food, water, shelter, and medical attention. That requires significant time, resources, and energy to ensure that animals are adequately provided for, and their welfare is safeguarded. Whether it's ensuring that pets are fed and exercised daily or that livestock have access to clean water and suitable living conditions, animal care responsibilities are ongoing and non-negotiable. Moreover, regular veterinary care and preventive measures, such as vaccinations and parasite control, are essential for responsible animal stewardship, requiring financial planning and commitment.

Furthermore, understanding the commitment and responsibilities of animal care extends beyond meeting their physical needs to encompass their emotional and psychological well-being. Animals, like humans, have complex social and emotional lives, and their mental health is just as important as their physical health. It means providing opportunities for enrichment, socialization, and mental stimulation, whether through interactive play for pets or access to outdoor pasture and companionship for livestock. Additionally, recognizing and addressing signs of distress or discomfort, such as behavioral changes or signs of illness, is crucial for maintaining the overall welfare of animals under one's care.

In addition to meeting the immediate needs of animals, responsible caretakers must also consider the long-term implications of their decisions and actions. Includes making informed choices about breeding, adoption, or acquisition of animals, considering factors such as population control, genetic diversity, and the capacity to provide lifelong care. Furthermore, responsible animal stewardship entails planning for the future and making provisions for the care of animals in the event of unforeseen circumstances, such as illness, relocation, or financial hardship. Creating backup plans, like pet trusts or agreements with reliable careers, may be necessary to guarantee that animals are cared for in an emergency.

Moreover, understanding the commitment and responsibilities of animal care extends beyond the individual level to encompass broader ethical and moral considerations. As sentient beings capable of experiencing pain, pleasure, and a range of emotions, animals deserve to be treated with dignity, respect, and compassion. It means rejecting practices that cause unnecessary suffering or exploitation, such as factory farming, animal testing, or the exotic pet trade, and advocating for policies and practices that promote animal welfare and rights. It also means being willing to make personal sacrifices and lifestyle changes to align one's actions with these principles, whether adopting a plant-based diet, supporting cruelty-free products, or advocating for stronger animal protection laws.

In conclusion, understanding the commitment and responsibilities of animal care is essential for ensuring the well-being and welfare of the creatures entrusted to our care. It requires a deep understanding of their physical, emotional, and social needs and a willingness to make personal sacrifices and ethical choices to uphold their rights and dignity. Whether caring for pets, livestock, or wildlife, responsible animal stewardship is a lifelong commitment that requires dedication, compassion, and a sense of duty to those who depend on us for their survival and well-being; by embracing these responsibilities with integrity and empathy, we not only fulfill our obligations as caretakers but also enrich our own lives through the bond and connection we share with the animals in our care.

An Overview of What to Expect

An overview of what to expect when embarking on the journey of raising chickens encompasses a diverse array of considerations, challenges, and rewards. For those venturing into poultry keeping for the first time, understanding the foundational aspects of chicken husbandry is crucial for success. From selecting breeds and designing coops to managing health, nutrition, and egg production, prospective poultry keepers must be prepared to navigate a myriad of factors to ensure the well-being and productivity of their flocks.

One of the initial considerations for aspiring chicken keepers is selecting the right breed of chickens for their specific needs and preferences. With hundreds of chicken breeds to choose from, each with its own unique characteristics, temperaments, and egg-laying abilities, it's essential to research and evaluate options carefully. Factors such as climate, available space, desired egg color and size, and suitability for backyard or small-scale production should all be considered when choosing breeds.

Once breeds have been selected, the next step is designing and constructing a suitable coop to provide shelter, security, and comfort for the flock. Coop design considerations include size, ventilation, insulation, predator protection, nesting boxes, roosting perches, and access to food and water. Whether building a coop from scratch or repurposing an existing structure, poultry keepers must ensure that the coop meets the specific needs of their flock and provides a safe and conducive environment for chickens to thrive.

Managing the health and well-being of chickens is another critical aspect of poultry keeping. Regular health checks, vaccinations, parasite control measures, and proper nutrition are essential for preventing diseases and ensuring that chickens remain healthy and productive. Monitoring chickens for signs of illness or injury, such as lethargy, decreased appetite, respiratory distress, or abnormal behavior, allows poultry keepers to intervene promptly and provide the necessary care and treatment.

Nutrition plays a crucial role in supporting chickens' overall health, immune function, and egg-laying performance. A balanced diet that provides the necessary nutrients, vitamins, and minerals is essential for maintaining chickens' health and productivity. Commercially formulated layer feeds serve as the foundation of chickens' diets, supplemented with fresh fruits, vegetables, grains, and protein sources such as mealworms or kitchen scraps to add variety and enrichment. Additionally, providing access to grit and oyster shell helps chickens digest their food properly and maintain strong, healthy eggshells.

Managing egg production is a central focus for many poultry keepers, as fresh eggs are often one of the primary motivations for raising chickens. Understanding the factors that influence egg production, such as breed, age, nutrition, daylight length, and environmental conditions, allows poultry keepers to optimize egg-laying performance and maximize egg yields. Monitoring chickens' laying patterns, egg quality, and overall health allows poultry keepers to identify any issues or disruptions to egg production and take appropriate measures to address them.

In addition to the practical aspects of chicken husbandry, prospective poultry keepers should also be prepared for the emotional and social aspects of raising chickens. Developing strong bonds with chickens, observing their behaviors, and experiencing the joys of caring for living creatures can be incredibly rewarding. However, poultry keeping also requires dedication, patience, and a willingness to adapt to the challenges and responsibilities that come with caring for living animals.

In conclusion, embarking on the journey of raising chickens is a multifaceted and rewarding endeavor that requires careful planning, dedication, and a commitment to learning. By understanding the foundational aspects of chicken husbandry, including breed selection, coop design, health management, nutrition, and egg production, prospective poultry keepers can set themselves up for success and enjoy the many benefits of chicken keeping. With proper care, attention, and a passion for poultry, raising chickens can be a fulfilling and enriching experience for individuals and families alike.

CHAPTER II

Choosing the Right Chicken Breed

Popular Chicken Breeds for Beginners

Understanding the fundamentals of chicken farming is a crucial first step for anyone venturing into the poultry-keeping hobby. This knowledge forms the bedrock for maintaining the health, nutrition, and productivity of your flock. From breed selection and coop building to health management, nutrition, and egg production, prospective chicken keepers must be prepared to handle a variety of issues.

Choosing the right breed of chicken is critical for prospective chicken keepers. With hundreds of breeds, each with its own unique traits, personalities, and egg-laying capacities, thorough research is a must. Factors such as climate, available space, desired egg size and color, and suitability for backyard or small-scale production should all be considered when making this important choice.

The next stage after choosing breeds is to plan and build an appropriate coop that will offer the flock comfort, security, and shelter. Size, ventilation, insulation, protection from predators, nesting boxes, roosting perches, and easy access to food and water are all factors in coop design. Poultry keepers need to ensure that their coop is safe, favorable to the growth of their flock, and fits the needs of their birds, whether they are building a new coop from the ground up or refurbishing an old one.

Maintaining the health and welfare of hens is a vital aspect of poultry keeping. Regular health examinations, immunizations, parasite control, and a balanced diet are all key to disease prevention and maintaining the health and productivity of hens. By being vigilant for signs of illness or injury and providing immediate care, poultry keepers can ensure the longevity and productivity of their flock.

Nutrition is essential to maintaining a chicken's general health, immune system, and ability to lay eggs. Maintaining hens' health and production requires a balanced diet rich in vitamins, minerals, and other nutrients. The basis of hens' meals is commercially prepared layer feed enriched and varied with fresh produce, grains, and protein sources like mealworms or leftover kitchen trash. Giving hens access to grit and oyster shells also aids in correct digestion and maintaining robust, healthy eggshells.

Since raising hens is frequently done primarily to produce fresh eggs, controlling egg production is a significant concern for many poultry keepers. Poultry keepers may maximize egg yields and optimize egg-laying efficiency by being aware of the variables that affect egg production —breed, age, nutrition, length of daylight, and environmental conditions. By keeping an eye on the laying patterns, egg quality, and general health of their hens, poultry caretakers can spot any problems or disturbances in egg production and take the necessary action to resolve them.

Prospective poultry keepers should be ready for the emotional and social components of rearing chickens and the practical aspects. It can be immensely fulfilling to form close relationships with hens, watch their behaviors, and enjoy the pleasures of taking care of living things. But maintaining chickens also calls for commitment, tolerance, and a readiness to accept the difficulties and obligations that come with taking care of living things.

In conclusion, learning to raise chickens is a joyful and diverse undertaking that requires careful preparation, perseverance, and a willingness to learn new things. Prospective poultry keepers can position themselves for success and reap the numerous rewards of chicken keeping by being aware of the industry's fundamentals, such as breed selection, coop design, health management, nutrition, and egg production. Raising chickens may be a rewarding and enjoyable experience for people and families if they receive the right care and attention and are passionate about poultry.

Factors to Consider When Selecting a Breed

Anyone starting a backyard poultry operation must make the crucial choice of choosing the correct breed of chickens. Selecting the best breeds can be challenging because hundreds of varieties have distinct qualities and attributes. However, people can limit their options and find breeds that suit their needs and tastes by carefully considering various criteria, including temperament, egg production, climate adaptation, space requirements, and intended purpose.

Temperament is one of the most important things to consider when choosing a breed. There is a vast spectrum in the temperaments of chickens, from gentle and amiable to more wary or hostile. Novices or families with young children must select breeds with a reputation for gentleness and calmness. An outgoing and gregarious breed is likelier to flourish in a backyard environment and get along well with their human caregivers. On the other side, more excellent handling and socialization may be necessary for breeds that are more aggressive or flighty to feel comfortable around people.

Egg production is another critical issue to consider. When choosing a breed of chicken to raise for fresh eggs, it's essential to look for those recognized for their prolific egg-laying ability. Certain breeds, such as Australorp, Leghorn, and Rhode Island Red, are known for their high egg production rates and year-round massive egg production. Conversely, breeds that serve two purposes, such as the Orpington and Plymouth Rock, are prized for balancing meat quality with egg output, making them appropriate for people who want to produce both meat and eggs.

The breed's capacity to adapt to different climates is another crucial factor. Although chickens are incredibly versatile, some are more adapted to particular regions and environments than others. Breeds that thrive in colder climes, like the Sussex and Wyandotte, are well-known for their ability to withstand harsh winters. On the other hand, Mediterranean breeds like the Leghorn and Minorca do well in hot conditions and are more adapted to warmer areas. People may guarantee the health and happiness of their flock all year round by selecting breeds that are well-suited to their particular climate.

When choosing a breed, space constraints are another thing to consider. Even though they are small animals, chickens still need enough room to live and grow. While some breeds require more room to wander and display their natural characteristics, others are better suited to confinement and do well in smaller coops and runs. For instance, bantam breeds are minor variations of regular chicken breeds that work well in urban or small-space settings. On the other hand, larger breeds, such as the Brahma and Jersey Giant, may not be appropriate for people with limited space since they need more room to spread their legs.

When choosing a breed, intended function is arguably the most crucial consideration. In addition to producing eggs and meat, chickens can also be used for pest control, decoration, display, and showmanship. People must consider their priorities and goals while selecting breeds that suit their intended use. People who are primarily concerned with producing eggs, for instance, might give preference to breeds that lay eggs frequently. In contrast, people more interested in producing meat might concentrate on breeds that produce meat at a higher grade and faster growth rate. Similar to this, people who are interested in ornaments could give preference to breeds with eye-catching feathers and distinctive physical traits.

When choosing a breed, people should also consider the availability of the breeds in their area. Certain breeds can be easily found at nearby hatcheries, while others might be harder to locate or need special orders. To find out whether desired breeds are available and to guarantee that you have access to healthy, premium birds, it is imperative to research local breeders and hatcheries. People should also consider the credibility and dependability of hatcheries and breeders to prevent problems with illness, genetic flaws, or subpar breeding procedures.

In conclusion, everyone considering raising chickens in their backyard must make the critical choice of choosing the proper breed. People can select breeds that suit their needs and preferences by carefully weighing characteristics, including temperament, egg production, climatic adaptation, space requirements, intended use, and availability. There are many possibilities to fit every lifestyle and inclination, whether you're looking for robust egg layers, cold-hardy breeds, space-saving options, or distinctive ornamental breeds. People can discover the ideal breeds to begin their chicken-keeping journey and take advantage of the many benefits of keeping hens by carefully researching and considering their options.

Matching Chicken Breeds with Your Needs and Environment

An essential step in raising poultry in your backyard is matching breeds of chickens with your needs and surroundings. Choosing the correct breeds of chickens can seem complicated because there are hundreds of varieties, each with distinctive qualities and attributes. However, people can narrow their options and find breeds well-suited to their particular needs and surroundings by carefully evaluating criteria including temperament, egg production, climatic adaptation, space requirements, intended use, and personal preferences.

Temperament is one of the most crucial aspects when matching hen breeds with your needs and surroundings. There is a vast spectrum in the temperaments of chickens, from gentle and amiable to more wary or hostile. Selecting breeds with a reputation for being gentle and peaceful is crucial for people searching for backyard companions or those with youngsters. Gregarious and friendly breeders have a higher chance of thriving in a backyard environment and getting along well with their human caregivers. Conversely, more aggressive or flighty breeds could need more care and socialization before they feel at ease with people.

The generation of eggs is another critical factor. When choosing a breed of chicken to raise for fresh eggs, it's essential to look for those recognized for their prolific egg-laying ability. Certain breeds, such as Australorp, Leghorn, and Rhode Island Red, are known for their high egg production rates and year-round massive egg production. Furthermore, breeds that serve multiple purposes, such as the Orpington and Plymouth Rock, are prized for balancing meat quality with egg output, making them appropriate for people who want to produce both meat and eggs.

Consideration should also be given to climate adaptability when choosing hen breeds for your needs and surroundings. Despite their remarkable adaptability, some varieties of chickens are more adapted to particular climates and environmental situations than others. Breeds that thrive in colder climes, like the Sussex and Wyandotte, are well-known for their ability to withstand harsh winters. On the other hand, Mediterranean breeds like the Leghorn and Minorca do well in hot conditions and are more adapted to warmer areas. You can guarantee the health and happiness of your flock all year round by selecting breeds that are well-suited to your area's climate.

While choosing hen breeds for your needs and surroundings, space limitations are another factor to consider. Even though they are small animals, chickens still need enough room to live and grow. While some breeds require more room to wander and display their natural characteristics, others are better suited to confinement and do well in smaller coops and runs. For instance, bantam breeds are minor variations of regular chicken breeds that work well in urban or small-space settings. On the other hand, larger breeds, such as the Brahma and Jersey Giant, may not be appropriate for people with limited space since they need more room to spread their legs.

The most crucial thing to remember when matching chicken breeds to your needs and surroundings is their intended use. In addition to producing eggs and meat, chickens can also be used for pest control, decoration, display, and showmanship. Selecting breeds for your intended use requires careful consideration of your priorities and specific aims. For instance, you might prefer breeds recognized for their prodigious egg-laying prowess if your primary concern is egg production. On the other hand, you might concentrate on breeds with exceptional meat quality and development rates if you're interested in producing meat.

When choosing a chicken breed for their needs and surroundings, people should also consider the availability of the breeds in their area. Certain breeds can be easily found at nearby hatcheries, while others might be harder to locate or need special orders. To find out whether desired breeds are available and to guarantee that you have access to healthy, premium birds, it is imperative to research local breeders and hatcheries. People should also consider the credibility and dependability of hatcheries and breeders to prevent problems with illness, genetic flaws, or subpar breeding procedures.

Once you've chosen the purpose of your backyard flock (eggs, meat, or show), breed temperament is an additional factor to consider. While the temperament characteristics of each breed are generally the same, every bird will have its personality. Consider looking at a more docile breed if you intend to keep a small backyard flock of three or five birds that will be loved as pets and contribute to the family's local food chain. If you intend to teach kids how to care for and maintain animals, pick a breed with traits that align with your family's values so the kids will look forward to the experience.

Breeds known for their docility, such as Wyandotte, Orpington, and Plymouth Rock, are not only suitable options for a beginner who wants to grow chickens in their backyard but also for families with little children who will be assisting with the raising and maintenance of the birds. By choosing these breeds, you can feel secure and comfortable in your choice, knowing that they are inherently gentle and safe for your family to interact with. However, it's important to note that roosters, regardless of breed, are inherently hostile, thus they might not be appropriate for rookie poultry enthusiasts or flocks where young people are tending to the birds. Individual birds' behavior is unpredictable. The breed's traits should serve as a guide, keeping in mind that temperamental selection is challenging. A reliable and experienced breeder is a beautiful place to start when deciding which breed of chickens to get for your backyard flock.

To sum up, choosing a breed of chicken that fits your needs and environment is an integral part of raising poultry in your backyard. People can select breeds ideal for their needs and preferences by carefully weighing characteristics, including temperament, egg production, climate adaptation, space requirements, intended use, and availability. There are many possibilities to fit every lifestyle and inclination, whether you're searching for amiable backyard companions, prolific egg layers, cold-hardy breeds, or unusual ornamental breeds. People can discover the ideal breeds to begin their chicken-keeping journey and take advantage of the many benefits of keeping hens by carefully researching and considering their options.

CHAPTER III

Setting Up Your Chicken Coop

Designing the Perfect Coop: Size, Structure, and Materials

One of the most important aspects of raising chickens in your backyard is designing the ideal coop, which will give your birds a secure, cozy, and helpful place to live. When designing a coop that satisfies the requirements of the hens and their caregivers, several things must be considered, from choosing the proper materials to figuring out the ideal size and structure. Size is the most crucial factor to consider. The chicken should have plenty of room to roost, nest, and comfortably walk around. As a general rule of thumb, give each chicken at least 2-3 square feet of indoor room, with extra space for the outside run area. It's important to err on the side of caution and give your hens plenty of space to spread their wings because overpopulation can cause stress, aggression, and health problems in the flock.

The coop's construction is as essential to its endurance and functionality as its size. Consider lighting, ventilation, and predator protection when building the coop. Maintaining enough ventilation is crucial to avoiding moisture buildup, which can cause respiratory problems and mold formation. Incorporate windows, vents, or other openings to ensure sufficient ventilation throughout the coop. Furthermore, ensure the coop has adequate lighting, either artificial or natural, to encourage wholesome egg production and deter hostile behavior from the flock. Last, use strong building materials and safe construction methods to keep foxes, rats, and raccoons out of the coop. To stop unwanted entrance, fortify weak spots and cover apertures with hardware cloth or welded wire mesh.

Durability and sustainability are important factors to consider when choosing materials. Select materials that won't rot in the weather are simple to clean and don't contain any harmful chemicals that could endanger your hens. Because of their endurance and durability, metal roofing, plywood, and pressure-treated lumber are common materials used in coop buildings. Alternatively, consider building your coop from repurposed or salvaged materials to lessen its environmental impact and give it personality. To ensure that your coop endures the test of time and offers your chickens a secure and comfortable place to live, put quality and craftsmanship above all else when selecting the materials.

Besides the coop's basic design, consider adding amenities that will improve its usability and convenience for you and your chickens. And maximize room and accessibility, features like feeders and waterers, roosting bars, and nesting boxes should be arranged thoughtfully within the coop. Create a cozy and secluded area for laying eggs; nesting boxes must be raised off the ground and coated with soft bedding. For suit hens of various sizes and tastes, roosting bars should be positioned at multiple heights. Incorporating hinged or detachable panels would provide simple access to the coop for upkeep and cleaning.

Remember to take your coop's aesthetic attractiveness into account. Even though durability and functionality are crucial, creating an aesthetically beautiful coop can increase how much pleasure raising chickens in your backyard can be. Add architectural details like window boxes, ornate trim, or a lovely color scheme to give your coop flair and charm. You and your hens can enjoy a pleasant and scenic environment by adding plants and flowers to the area surrounding the coop.

Creating a safe, cozy, and helpful living area for your hens necessitates careful consideration of size, construction, and materials when planning the ideal coop. You can ensure that your coop serves the needs of your chickens and their caregivers by prioritizing it when it comes to enough room, appropriate ventilation, robust construction, and sustainable materials. Furthermore, adding amenities that improve accessibility, usefulness, and aesthetic appeal can improve backyard poultry farming and create a quaint and cozy area you and your flock can enjoy.

The health, safety, and well-being of the flock are all factors that must be carefully considered while designing the ideal chicken coop. These characteristics include the coop's dimensions, design, amenities, security, and ventilation. Prioritizing these fundamental components will help poultry keepers build a cozy and practical environment that caters to their birds' natural needs and behaviors.

Given that it directly affects the flock's comfort and welfare, size is essential in coop design. For a given number of hens, a coop should have enough room to walk freely and engage in natural activities like scratching, foraging, and dust bathing. Generally speaking, provide each chicken at least 2-3 square feet of inside space, plus extra space for movement and exploration in the outdoor run area. To prevent stress, aggression, and health problems in the flock, providing enough room for the hens' needs is crucial.

To sum up, building a chicken coop requires all the necessary components to give the flock a secure, cozy, and valuable living area. Assure the health and welfare of the hens by giving them enough room, adequate ventilation, security measures, nesting boxes, and roosting bars priority. To improve your experience with backyard poultry husbandry overall, plan the coop with ease, functionality, and aesthetics in mind. Your flock will have a happy and healthy home if you build a coop that suits the needs of your caregivers and your hens with careful planning and attention to detail.

Essential Features of a Chicken Coop

Essential features of a chicken coop are fundamental to providing a safe, comfortable, and functional living environment for chickens. A well-designed coop protects chickens from predators, harsh weather, and disease and promotes natural behaviors, such as roosting, nesting, and foraging. Whether you're a beginner or an experienced poultry keeper, understanding the essential features of a chicken coop is crucial for successful chicken raising.

First and foremost, adequate space is essential for a chicken coop. Chickens require enough room to move around freely, roost comfortably, and engage in natural behaviors such as dust bathing and foraging. As a general rule of thumb, allow at least 2-4 square feet of floor space per chicken inside the coop, with additional space provided in the outdoor run. Overcrowding can lead to stress, aggression, and disease, so it's essential to ensure that the coop is spacious enough to accommodate the flock's size comfortably.

Ventilation is another critical feature of a chicken coop. Good airflow helps regulate temperature and humidity levels inside the coop, preventing moisture buildup, ammonia odors, and respiratory issues. Vents or windows should be installed strategically to allow cross-ventilation while protecting chickens from drafts and predators. Adjustable vents or windows are ideal, allowing airflow control based on weather conditions and seasonal changes.

Protection from predators is paramount in a chicken coop. Predators such as foxes, raccoons, snakes, and birds of prey pose a significant threat to chickens, especially at night. Deter predators; the coop should be constructed with sturdy materials and reinforced with hardware cloth or wire mesh to prevent entry. To prevent unauthorized access, secure locks and latches should be used on doors and windows. Additionally, burying wire mesh around the perimeter of the coop helps prevent burrowing predators from gaining access from below.

Nesting boxes are essential features of a chicken coop. They provide a designated space for chickens to lay their eggs. Nesting boxes should be dark, quiet, and private, mimicking the natural environment where hens typically lay their eggs. Provide one nesting box for every 3-4 hens, ensuring each box is large enough to accommodate a hen comfortably. The boxes should be lined with clean bedding material, such as straw, hay, or wood shavings, to provide a comfortable and hygienic nesting environment.

Roosting bars are another essential feature of a chicken coop. They provide chickens with a designated space to perch and sleep at night. Roosting bars should be installed at least 18 inches above the ground and spaced 8-12 inches apart to accommodate chickens of all sizes. The bars should be made of smooth, rounded materials, such as wooden dowels or PVC pipes, to prevent injury to chickens' feet. Multiple roosting bars at varying heights allow chickens to establish a pecking order and choose their preferred roosting spot.

Flooring material is an essential consideration in a chicken coop. The coop floor should be easy to clean, non-toxic, and resistant to moisture and odors. Common flooring materials include concrete, dirt, gravel, or wooden boards, depending on personal preference and climate conditions. Adding a layer of absorbent bedding material, such as straw, hay, or wood shavings, helps absorb moisture, control odors, and provide a comfortable surface for chickens to walk on.

Access to food and water is essential for chickens' health and well-being. The coop should have sturdy feeders and waterers resistant to tipping, spilling, and contamination. To prevent contamination by droppings, position feeders and waterers away from nesting boxes and roosting areas. Use hanging or wall-mounted feeders and waterers to conserve space and avoid spillage. Additionally, provide access to grit and oyster shells to help chickens digest their food correctly and maintain solid and healthy eggshells.

Lighting is essential in a chicken coop, particularly for promoting egg production and regulating chickens' circadian rhythms. Natural daylight is ideal, so position the coop in a location that receives ample sunlight throughout the day. Supplemental lighting can extend daylight during the shorter days of fall and winter, stimulating egg production and preventing disruptions to the flock's reproductive cycle. Timers can be used to automate lighting schedules, ensuring that chickens receive the appropriate amount of light each day.

Lastly, cleanliness and hygiene are essential to maintaining a healthy chicken coop. Regularly clean and disinfect the coop to remove droppings, dirt, and debris, reducing the risk of disease and parasites. Replace bedding material to keep the coop clean, dry, and odor-free. Inspect the coop regularly for signs of wear and tear, damage, or potential entry points for predators, and make repairs or reinforcements as necessary.

In conclusion, the essential features of a chicken coop are crucial for providing a safe, comfortable, and functional living environment for chickens. Adequate space, ventilation, protection from predators, nesting boxes, roosting bars, flooring material, access to food and water, lighting, and cleanliness are all considerations in coop design and management. By understanding and implementing these essential features, poultry keepers can create a conducive environment for their flocks to thrive and enjoy the many benefits of chicken raising.

Tips for Building or Buying a Coop

Building or buying a coop is a significant decision for anyone considering raising chickens. Whether you're a seasoned chicken keeper or a novice enthusiast, careful consideration of various factors is essential to ensure that the coop meets the needs of the chickens and their caretakers. From design and construction to location and cost, here are some tips to help you navigate the process of building or buying a coop.

First and foremost, consider your specific needs and requirements when deciding whether to build or buy a coop. Building a coop allows for customization and flexibility to design a coop that meets your specifications and preferences. You can tailor the coop's size, layout, and features to accommodate the number of chickens you plan to keep and the available space in your backyard. Building a coop can also be a rewarding DIY project for those who enjoy woodworking and construction. However, building a coop requires time, effort, and a certain skill level, so assessing your abilities and resources is essential before embarking on a DIY coop project.

On the other hand, buying a coop offers convenience and ease of installation, especially for those needing more time or expertise to build a coop from scratch.

Prefabricated coops are available in various sizes and designs to suit different needs and preferences, making it easy to find a coop that meets your requirements. Additionally, buying a coop may be a more cost-effective option for some, especially if you factor in the time and materials required to build a coop yourself. However, it's essential to carefully research different coop options to ensure you're getting a quality product that meets your needs and budget.

When building or buying a coop, it's crucial to consider the size and layout of the coop to provide adequate space for your flock. Allow for at least 2-3 square feet of indoor space per chicken, with additional room in the outdoor run area for exercise and foraging. Overcrowding can lead to stress, aggression, and health issues among the flock, so it's essential to prioritize ample space for your chickens to move around comfortably. Additionally, consider the coop layout, including the placement of nesting boxes, roosting bars, and feeders and waterers, to optimize space and functionality.

Location is another critical consideration when building or buying a coop. Choose a location that is well-drained, level, and free from hazards such as flooding, predators, and excessive noise or disturbances. Ideally, the coop should be sunny, with access to natural light and ventilation to promote a healthy environment for the chickens. Additionally, consider factors such as proximity to your home, ease of access for cleaning and maintenance, and compatibility with local zoning regulations and ordinances governing backyard poultry keeping.

When building or buying a coop, prioritize durability and quality to ensure that the coop stands the test of time and provides a safe and comfortable living space for your chickens. Use sturdy materials and secure construction techniques to protect the coop from predators and the elements. Also, choose weather-resistant materials, easy to clean and free from toxic chemicals that could harm your chickens. Investing in a high-quality coop may require a more considerable upfront investment, but it can save you time and money in the long run by reducing the need for repairs and replacements.

Consider the long-term maintenance and upkeep of the coop when making your decision. Building or buying a coop is just the beginning – regular cleaning, maintenance, and repairs are essential to keep the coop in good condition and ensure the health and well-being of your flock. Consider the time, expense, and effort involved in continuous upkeep when selecting a coop design or product. Additionally, consider the availability of resources such as feed, bedding, and medical supplies, as well as the support and advice of other chicken keepers in your community, to help you care for your flock effectively.

Finally, consider the overall cost of building or buying a coop, including upfront and ongoing maintenance costs. Compare the cost of materials and labor required to build a coop yourself with the price of prefabricated coops available for purchase. Aside from that, budget for additional features or accessories, like feeders, waterers, roosting bars, and nesting boxes. While budget considerations are important, it's essential not to sacrifice quality or functionality to save money. Invest in a coop that meets your needs and provides your flock with a safe and comfortable home, even if it means spending more upfront.

In conclusion, building or buying a coop is a significant decision that requires careful consideration of various factors. Whether you build a coop yourself or purchase a prefabricated coop, prioritize ample space, durability, quality, and functionality to ensure that the coop meets the needs of your chickens and their caretakers. Consider factors such as size, layout, location, maintenance, and cost when deciding, and seek advice from other chicken keepers or poultry experts to help you choose the best coop for your needs and budget. With careful planning and consideration, you can build or buy a coop that provides a safe, comfortable, and functional home for your flock, allowing you to enjoy the many rewards of backyard poultry keeping.

CHAPTER IV

Creating a Safe and Comfortable Run

Planning and Designing a Chicken Run

Creating and planning a chicken run is an essential part of raising poultry in your backyard. It provides a safe and secure outdoor space for chickens to wander, forage, and engage in their natural activities. Thorough planning and consideration of important variables are necessary to create a useful and welcoming habitat for your flock, whether you're beginning from scratch or renovating an existing area.

Size is one of the first factors to consider while organizing and creating a chicken run. For the number of chickens, you intend to maintain, the run should have enough room to roam around and partake in their natural activities, such as pecking, dust bathing, and scratching. Provide each chicken at least 10 square feet of outside space as a general rule of thumb, though more space is always preferable. Remember that hens are curious and energetic creatures, so give them room to roam around to support their physical and emotional health.

Security is crucial when building a chicken run to keep the flock safe from predators. Whether creating a run from the ground up or remodeling an existing area, ensure it's built with durable materials and safe building practices to keep away animals like foxes, dogs, and raccoons. Cover the entire run's perimeter, including the roof, with hardware cloth or welded wire mesh that has tiny gaps to prevent unwanted access. To keep predators from digging, bury the wire mesh underground or cover it with an apron around the perimeter.

To protect the flock from the weather and give them a comfortable habitat, the design of the chicken run must include shade and shelter. Consider adding a solid roof or awning over a section to preserve the run from the rain, snow, and direct sunshine. Use objects like trees, plants, or umbrellas to create shady spots within the run. To avoid heat stress during hot weather, ensure the covered areas are well-ventilated and drafts-free.

When organizing and creating a chicken run, accessibility and practicality are crucial factors. Ensure the run is conveniently accessible for upkeep, cleaning, and egg collection. Consider including entrances or gates in the design to make it easy to enter and depart the run. To guarantee that the flock has access to food and water at all times, offer them access to both inside the run. Add automatic feeders and waterers to reduce daily maintenance duties and ensure the flock is well-fed.

To enhance the physical and mental health of the flock, consider designing the chicken run with enrichment features in addition to shelter and shade. To promote natural behaviors in your hens, provide them with structures like dust baths and pecking blocks, as well as perches, logs, and branches for them to explore and roost on. Add greenery, such as grass, herbs, and edible plants, to the run to give the flock more places to forage and to become excited.

While building and planning a chicken run, it is imperative to consider the surrounding environment and potential hazards that could endanger the flock. Make sure the run is situated far from any possible risks, such as busy roads and predators' habitats. Furthermore, eliminate any poisonous plants or materials from the vicinity of the run to avoid the flock accidentally consuming them. To keep off intruders who might endanger the flock's safety, including stray dogs or neighboring pets, consider erecting a fence or other barriers.

Lastly, pay attention to the chicken run's visual charm. Though practicality and utility come first, creating an aesthetically beautiful rug can add to the overall experience of raising chickens in your garden. Other elements that can be used to create a cozy and enticing environment for you and your hens to enjoy including flower beds, mulched areas, and decorative pieces.

To sum up, creating a safe, secure, and engaging outdoor environment for your hens requires careful consideration of several important aspects when planning and developing a chicken run. For your hens' health and welfare, prioritize enough room, safety precautions, cover, shade, accessibility, ease of use, enrichment elements, and environmental concerns. You may build a hen run that satisfies the requirements of your flock and their caregivers with careful planning and attention to detail, giving your hens a contented and healthy outdoor habitat to grow.

Ensuring Security from Predators

Ensuring security from predators is a paramount concern for backyard poultry keepers, as predators pose a significant threat to the safety and well-being of chickens. Whether you're raising chickens in a suburban backyard or on a rural farm, it's essential to take proactive measures to protect your flock from potential predators such as raccoons, foxes, coyotes, birds of prey, snakes, and domestic pets. By implementing a combination of physical barriers, deterrents, and proactive management strategies, you can create a safe and secure environment for your chickens to thrive.

One of the most effective ways to protect your chickens from predators is to secure their coop and run with sturdy fencing and predator-proofing measures. Use materials such as hardware cloth, welded wire mesh, or chain-link fencing to create a barrier around the perimeter of the coop and run, burying the fencing underground to prevent digging predators from gaining access. Ensure that the fencing is tall enough to prevent climbing predators from scaling it and consider adding a protective skirt or apron along the bottom of the fence to deter digging. Additionally, reinforce weak points such as doors, windows, and vents with heavy-duty locks, latches, and hardware to prevent unauthorized entry.

Another essential aspect of predator security is securing the coop with sturdy construction and predator-proofing features. Ensure the coop is built with durable materials and secure construction techniques to withstand attempted predator breaches. Use heavy-duty hardware cloth or welded wire mesh to cover openings such as windows, vents, and wall gaps and reinforce weak points with additional framing or bracing. Consider installing predator-proof latches on doors and windows and motion-activated lights or alarms to deter nocturnal predators such as raccoons and foxes.

In addition to physical barriers, consider implementing deterrents and proactive management strategies to discourage predators from targeting your flock. Install motion-activated sprinklers or sound devices around the perimeter of the coop and run to startle and deter predators when they approach. Use natural deterrents such as predator urine, predator decoys, or reflective surfaces to create the illusion of a more prominent, more formidable predator presence in the area. Additionally, practice good husbandry practices such as promptly removing food and water sources from the coop at night, securing trash and compost bins, and avoiding leaving pet food outside, which can attract predators to your property.

When designing your coop and run, consider incorporating additional security features such as electric fencing or predator-proof roofing to deter predators further and protect your flock. Electric fencing can effectively deter larger predators such as coyotes and bears, delivering a mild shock to discourage them from attempting to breach the enclosure. Similarly, predator-proof roofing from sturdy materials such as metal or corrugated plastic can prevent aerial predators such as hawks and owls from swooping down and attacking chickens from above.

Finally, remain vigilant and proactive in monitoring and managing predator activity around your property. Regularly inspect the perimeter of the coop and run for signs of attempted breaches or damage, such as scratches, scuff marks, or paw prints. Keep an eye out for any unusual behavior or signs of disturbance among your flock, such as nervousness, agitation, or injuries, which may indicate predator activity in the area. By staying alert and promptly addressing potential threats, you can help keep your chickens safe and secure from predators.

In conclusion, ensuring security from predators is critical to backyard poultry keeping, requiring proactive measures to protect your flock from potential threats. By implementing a combination of physical barriers, deterrents, and proactive management strategies, you can create a safe and secure environment for your chickens to thrive. From securing the coop and run with sturdy fencing and predator-proofing measures to implementing deterrents and proactive management practices, there are numerous steps you can take to safeguard your flock from predators and provide them with a safe and comfortable living environment. With careful planning, vigilance, and proactive management, you can help keep your chickens safe and secure from predators and enjoy the rewards of backyard poultry keeping for years to come.

Providing Space and Enrichment for Your Chickens

Providing space and enrichment for your chickens is essential for their physical and mental well-being and ensuring that they lead happy and healthy lives. Like all animals, chickens have natural behaviors and instincts that need to be fulfilled, and it's the responsibility of poultry keepers to create an environment that allows them to express these behaviors. One of the most critical aspects of chicken farming is providing adequate space for the flock to move around and engage in natural behaviors. As a general rule of thumb, provide at least 2- 3 square feet of indoor space per chicken in the coop, with additional space in the outdoor run area. However, more space is always better, as overcrowding can lead to stress, aggression, and health issues among the flock. Allow chickens plenty of room to roam, scratch, peck, dust, bathe, and forage, as these activities are essential for their physical and mental well-being.

In addition to providing ample space, enriching the environment is essential to keeping chickens mentally stimulated and engaged. Chickens are highly intelligent and curious creatures that thrive on mental stimulation, so providing various enrichment activities can help prevent boredom and improve their overall welfare. Consider incorporating perches, logs, stumps, and rocks for chickens to climb on and explore. These structures provide exercise and mental stimulation, mimic natural environments, and encourage chickens to engage in natural behaviors such as roosting and perching.

Enrichment is often thought of as an extra or optional provision for residents. Sanctuary workers are understandably focused on providing the food, water, and housing necessary for residents to live. However, we hope that by incorporating enrichment as an aspect of general care, the lives of residents will be enriched. A particular importance for residents residing in smaller, more confined, or barren living spaces. In areas that experience intense cold, the best way to keep chicken residents warm is to leave them in a smaller, coop-like space when they may generally have a larger outdoor space for foraging and other chicken behaviors.

Another enriching activity for chickens is foraging for food. Scatter scratch grains, seeds, or mealworms throughout the coop and run area to encourage chickens to search for food and peck at the ground. Additionally, consider planting a diverse range of vegetation, such as grass, clover, herbs, and vegetables, in the outdoor run area to provide chickens with fresh greens and encourage natural foraging behaviors. Chicken-friendly plants provide nutritional benefits and offer shade, shelter, and opportunities for chickens to explore and discover new things.

Dust bathing is another essential behavior for chickens, as it helps keep their feathers clean and healthy and controls parasites such as mites and lice. Provide a designated dust bathing area within the coop or run, filled with loose soil, sand, or diatomaceous earth, where chickens can indulge in their instinct to dust bath. Ensure that the dust bathing area is kept dry and replenish the substrate regularly to keep it clean and inviting for the flock.

Water access is crucial for chickens, as they always require clean, fresh water to stay hydrated and healthy. Provide multiple water sources throughout the coop and run area, ensuring that chickens always have easy access to water. Consider using waterers with nipples or cups to prevent contamination and minimize spillage, especially in outdoor environments where water can quickly become dirty or contaminated. Additionally, regularly clean and refill waterers to ensure that chickens always have access to fresh water.

In conclusion, providing space and enrichment for your chickens is essential for their physical and mental well-being and ensuring they lead happy and healthy lives. Prioritize ample space, enrichment activities, foraging opportunities, dust bathing areas, and clean water sources to create an environment that allows chickens to express natural behaviors and thrive. By enriching the environment and meeting the needs of your flock, you can ensure that your chickens are happy, healthy, and contented members of your backyard flock.

CHAPTER V

Bringing Your Chickens Home

Preparing for the Arrival of Your Chickens

Preparing for the arrival of your chickens is an exciting and crucial step in your journey as a chicken owner. Whether you start with a small flock of backyard hens or plan to raise chickens for eggs or meat production, proper preparation ensures a smooth transition for both you and your feathered friends.

First and foremost, before your chickens arrive, it's essential to have a suitable living space ready for them. This includes a well-designed and adequately sized chicken coop or housing structure where your chickens will roost, nest, and seek shelter. The coop should be predator-proof, providing protection from predators such as foxes, raccoons, and birds of prey. Additionally, ensure that the coop is well-ventilated, with adequate airflow to prevent moisture buildup and maintain optimal air quality for your chickens.

Chickens are happiest with adequate outdoor space to roam. A run attaches to your coop and should provide at least 10 square feet (0.9 square meters) of space per bird. Again, this is a general recommendation. The more space you can provide, the better. Of course, if you don't have a lot of predators in your area, you can forgo the run and let them free range — meaning letting them explore your property freely. However, they still need to have access to the coop so that they can lay and shelter from predators at night. You'll also want dry dirt where your chickens can dust, bathe, or roll around in mud. This natural behavior helps keep your chickens healthy.

In addition to a safe coop and space to roam, chickens need feed, water, and a source of calcium. Use Scratch and Peck feed for our chicks, pullets, and hens. Chickens have different nutrient needs depending on age, so purchase the appropriate feed. Throw our feed on the ground, but you can buy a poultry feeder. Chickens need consistent access to clean drinking water. If you live in a cold climate, you must invest in a heated poultry drinker to keep the water from freezing in the winter.

Laying hens also need access to calcium to maintain healthy bones and lay eggs with strong shells. It would help if you gave your hens crushed oyster shells regularly. Furthermore, chickens confined to a coop or need access to grit. As chickens can't break down their food independently, they rely on grit — or small pieces of rock and stone — to break down their food in their gizzard.

In addition to the coop, you'll also need to set up a secure outdoor area where your chickens can roam and forage during the day. It can be a fenced-in chicken run attached to the coop or a designated area within your backyard enclosed with chicken wire or other suitable fencing materials. Providing ample space for your chickens to explore and engage in natural behaviors like scratching, pecking, and dust bathing is essential for their physical and psychological well-being.

Once the housing setup is in place, it's time to consider the necessary supplies and equipment for your chickens' care. Includes feeders, waterers, nesting boxes, bedding materials, and other accessories needed to create a comfortable and functional living environment. Depending on your preferences and budget, you may opt for essential, no-frills equipment or invest in more advanced, automated systems for feeding and watering.

Next, it's crucial to stock up on high-quality chicken feed formulated for the specific needs of your flock. Different breeds and ages of chickens may require different types of feed, so be sure to choose a suitable option based on your chickens' age, breed, and intended purpose (e.g., laying hens, meat birds, etc.). Additionally, consider offering supplemental treats and fresh fruits and vegetables to provide variety in your chickens' diet and keep them healthy and happy.

In addition to food and water, remember to gather essential health and first aid supplies for your chickens. These include poultry vitamins and supplements, poultry dust for mite and lice control, and a basic first aid kit containing supplies for treating common injuries and ailments. Being prepared for potential health issues and emergencies will enable you to respond quickly and effectively to keep your chickens healthy and thriving.
Furthermore, before bringing your chickens home, it's essential to familiarize yourself with local regulations and ordinances regarding chicken keeping in your area. Some municipalities may restrict the number of chickens you can keep and rules governing coop placement, noise ordinances, and other considerations. By understanding and adhering to these regulations, you can avoid potential conflicts and ensure a positive experience for you and your neighbors.

Lastly, preparing for the arrival of your chickens involves more than just physical preparation—it also requires mental and emotional readiness. Raising chickens is a commitment that requires time, effort, and dedication, so it's essential to approach it with a positive attitude and a willingness to learn. Take the time to educate yourself about chicken care and farming practices, seek advice from experienced chicken keepers, and be open to adapting and evolving your approach as you gain experience.

In conclusion, preparing for the arrival of your chickens is a multifaceted process that involves careful planning, preparation, and consideration of your chickens' needs.

By ensuring that you have a suitable living space, necessary supplies and equipment, and a solid understanding of chicken care basics, you can set yourself and your flock up for success from the start. With proper preparation and a commitment to providing the best possible care for your chickens, you'll be well on your way to enjoying the many rewards of chicken keeping.

Introducing Chickens to Their New Environment

Introducing chickens to their new environment is critical in backyard chicken keeping. Whether you are bringing home newly hatched chicks or mature birds, the transition to a new living space can be stressful for chickens. Therefore, careful planning and consideration are essential to ensure a smooth and successful integration into their new surroundings.

One of the first considerations when introducing chickens to their new environment is the physical setup of their living space. Includes providing a suitable coop or housing structure that offers protection from predators, shelter from the elements, and ample space for roosting, nesting, and foraging. The coop should be clean, well-ventilated, and free from potential hazards or obstacles that threaten the chickens' safety and well-being.

In addition to the coop, it's essential to establish a designated outdoor area for the chickens to roam and explore. This area can be a chicken run or fenced-in yard, providing the chickens with fresh air, sunlight, opportunities for exercise, and natural behaviors. Ensuring that the outdoor space is secure and predator-proofed is essential for the safety and security of the flock.

Before introducing chickens to their new environment, it's crucial to thoroughly clean and disinfect the coop and surrounding area to prevent the spread of disease and parasites. Remove debris, old bedding, or droppings, and sanitize surfaces with a poultry-safe disinfectant. Providing fresh bedding and nesting material will help create a clean and comfortable environment for the chickens.

When bringing home new chickens, whether chicks or mature birds, it's essential to provide them with a period of adjustment and acclimatization to their new surroundings. That may involve keeping them confined to the coop and running initially to allow them to become familiar with their new environment and establish a sense of security. Gradually introduce them to the outdoor space, supervised at first, to prevent them from becoming disoriented or wandering off.

During the initial introduction period, observe the chickens closely for signs of stress or aggression. Chickens are social animals that establish a pecking order within the flock, and introductions can sometimes lead to quarrels or territorial disputes. It's important to intervene if aggressive behavior occurs, separating the chickens if necessary and providing additional space or resources to reduce competition.

In addition to physical adjustments, it's essential to consider the social dynamics of introducing new chickens to an existing flock. If you already have chickens, introducing newcomers can disrupt the established pecking order and lead to conflicts. To minimize stress and aggression, introduce new chickens gradually, starting with short, supervised visits and progressively increasing the duration and frequency of interactions.

Providing multiple food and water sources and plenty of space for the chickens to retreat and establish their own territories can help reduce competition and tension within the flock. Monitor the chickens closely during the integration process, stepping in to intervene if any aggressive behavior occurs or if any chicken appears excessively stressed or isolated.

Furthermore, it is essential to ensure that the chickens' nutritional needs are met during the transition period. A balanced diet rich in protein, vitamins, and minerals will support their health and well-being as they adjust to their new environment. Additionally, offering treats and enrichment activities can help alleviate stress and boredom, promoting a sense of comfort and security.

As the chickens become more accustomed to their new environment and each other, they gradually expand their access to the outdoor space, allowing them to explore and forage freely. With time and patience, they will establish their routines and hierarchies within the flock, and the integration process will become smoother and more harmonious.

In conclusion, introducing chickens to their new environment requires careful planning, observation, and patience. By providing suitable living space, facilitating gradual introductions, monitoring social dynamics, and ensuring that their nutritional needs are met, you can help ease the transition and promote successful integration into their new surroundings. Ultimately, your chickens will thrive in their new home with proper care and attention, rewarding you with fresh eggs, companionship, and endless entertainment.

The First Few Days: What to Watch For

The initial days following the arrival of your chickens are a critical time for observation and adjustment. As your feathered friends settle into their new environment, it's not just a task, but a joy to pay close attention to their behavior, health, and well-being. This ensures a smooth transition and allows you to address any issues that may arise. From monitoring their eating and drinking habits to observing their social interactions within the flock, being vigilant during the first few days will set the stage for a successful and fulfilling chicken-raising experience.

One of the first things to watch for in the first few days is how well your chickens are adapting to their new surroundings. Moving to a new environment can be stressful for chickens, particularly if they are coming from a different flock or living situation. Your role is crucial in providing a calm and quiet environment, as well as plenty of space and opportunities for enrichment. This can help ease their transition and promote a sense of security, making them feel safe and cared for.

Additionally, monitoring your chickens' eating and drinking habits is essential during the first few days. Changes in environment or routine can sometimes cause chickens to lose their appetite or become dehydrated, quickly leading to health problems if left unaddressed. Ensure that your chickens have access to fresh water and offer high-quality feed appropriate for their age and breed. Monitor their food and water consumption and consult a veterinarian if you notice any significant changes or concerns.

Another essential aspect to watch for in the first few days is the establishment of social hierarchies within the flock. Chickens are social animals with complex social structures, and it's normal for them to engage in behaviors such as pecking, chasing, and vocalizations as they establish their place within the pecking order. While some level of aggression and conflict is expected during this time, it's essential to intervene if these behaviors become excessive or violent. Provide plenty of space and resources to minimize competition reduce stress and monitor their interactions closely to ensure a peaceful coexistence within the flock.

Furthermore, monitoring it's your chickens' health and well-being closely during the first few days is crucial. Stressful situations such as moving to a new environment or integrating into a new flock can compromise the immune system and increase the risk of illness. Look for signs of disease or injury, such as lethargy, decreased appetite, abnormal droppings, or respiratory symptoms, and consult a veterinarian if you have any concerns. Additionally, conduct regular health checks, inspect your chickens for external parasites or abnormalities, and provide appropriate preventive care.

Before the incubation calendar hits day 21, make sure you've ticked these tasks off your list. Otherwise, baby chickens may be in some strife early on! By now, you should have your brooder set up and the heat lamp warming to the right temperature. In that way, chickens will have ideal living conditions immediately instead of shivering while the heat lamp is warming up! The base of the brooder needs to be lined with bedding so it's easy to dispose of the baby chicken's droppings and so it helps the baby chickens build up their leg strength. For the first week or two, you should put thick strips of paper towels down as bedding. Once the chickens have had a couple of weeks to build up their leg strength, change it to shreds.

Baby chickens should have their first meal prepared and served in specially made baby chicken drinkers and feeders. You may be wondering why you can't use an ordinary container—well, baby chickens are small and can unfortunately fall into a container if it's too large. This puts them at risk of drowning and illness from soiling their food. Baby chicken drinkers and feeders have small outlets to access the water and feed, so they can't accidentally fall in. It's very, very tempting to help baby chickens hatch out of their shell. They look so helpless and worn out that we wish to break the shell ourselves! However, you definitely should not help your baby chicken hatch out of its eggshell, as many essential things take place that are crucial to their healthy development.

Baby chicken gets essential nutrients from eating parts of their eggshell, which is one of the reasons why you shouldn't discard it immediately! Also, the struggle of breaking apart the eggshell strengthens the baby chicken's muscles, which sets it up to have a healthy body. So, however tempting it may feel, don't help your baby chicken out of the shell - you'll cause them a wealth of health problems in the future. When your baby chickens hatch, their feathers might look a little damp. It might be tempting to move them into the warmth of the brooder, but you should only move the baby chickens once they've fluffed up. If you move them beforehand, they can catch a chill and become sick.

Baby chickens are marvelous little creatures. They cheer each other on, encouraging their brothers and sisters to hatch through cheeping! So, if you've had one chicken hatch, but it looks like more are on the way, don't take them out! They're helping more chickens hatch through vocal moral support. Within the first 24 hours of a baby chicken's life, they should have their first drink of water. Remember, the baby chickens have just hatched and don't know how to do anything except cheap and stumble around!

So don't be surprised if you put the waterer in the incubator, and they need to know what it is and how to drink it! The baby chickens need a little encouragement when learning how to drink. So, gently dip their beaks into the water if they still need to get the hang of it. They'll soon get the idea! The same goes for feeding. The baby chickens might need to learn how to eat the delicious starter feed in front of them. Again, gently dipping their head into the feed will coax them to start eating.

Your brooder temperature must be close to the mark—baby chickens need their house to be toasty warm in their first few weeks of life. Keep an eye on the heat lamp temperature; it should be 37.5 degrees, at least for the first week. Decrease it by about 1 degree every week after that before your chickens are ready to face the temperatures of the outside world! Mentioned earlier that you should use strips of paper towels as bedding. It's essential for the health of your baby chickens that you change the bedding regularly when it's soiled—you'll see that it doesn't take long to become dirty! Replace periodically with fresh, clean paper towel strips; your baby chickens will stay healthy and grow well.

In addition to physical health, it's essential to consider your chickens' mental and emotional well-being during the first few days. Chickens are intelligent and social creatures that thrive on routine and social interaction, and sudden changes or disruptions to their environment can cause stress and anxiety. Spend time observing your chickens' behavior and interactions and provide opportunities for enrichment and stimulation to keep them engaged and content. Include providing perches, toys, and opportunities for foraging and dust bathing, as well as interacting with them and building a bond based on trust and mutual respect.

Lastly, it is essential to be proactive and prepared to address any unexpected challenges or issues that may arise during the first few days. From minor adjustments to the living environment to more significant health concerns, your proactive approach is vital to ensuring their well-being and happiness. By educating yourself about common chicken health issues and basic first aid procedures and having a plan in place for seeking veterinary care if necessary, you can navigate the transition to your new environment with confidence and ease, feeling empowered in your role as a caretaker.

In conclusion, the first few days following your chickens' arrival are critical for observation and adjustment. By closely monitoring their behavior, health, and well-being, you can ensure a smooth transition and address any issues that may arise promptly. From providing a calm and secure environment to monitoring their eating and drinking habits and fostering positive social interactions within the flock, being vigilant and proactive during this time will set the stage for a successful and fulfilling chicken-raising experience.

CHAPTER VI

Nutrition and Feeding

Understanding Chicken Dietary Needs

Everything that an animal or plant does to take in and use its food is what we term nutrition. For any animal to carry on its necessary life functions, as well as to grow and be able to produce food for humans, there are specific chemical compounds that must be in their diet in the right amounts. No animal is capable of making every one of the necessary components on its own.

The study of nutrition is about understanding that there is no "perfect" food or no one ingredient that contains every nutrient an animal needs in the exact concentrations it needs. Providing the proper nutrition for your chickens means ensuring that what they eat supplies all the essential amino acids, fatty acids, carbohydrates, vitamins, minerals, and water needed to produce the meat or eggs you hope to collect.

Proper nutrition is essential for chickens' health, well-being, and productivity. As omnivores, chickens require a balanced diet that provides vital nutrients, vitamins, and minerals to support growth, egg production, and overall health. Understanding chickens' dietary needs is crucial for chicken keepers to ensure that their feathered friends receive the nutrition they need to thrive.

The foundation of a chicken's diet consists of high-quality commercial poultry feed explicitly formulated for their nutritional requirements. Poultry feed typically comes in pellets, crumbles, or mash and is available in various formulations tailored to different life stages and purposes, such as starter, grower, layer, and broiler feeds. These feeds are carefully formulated to provide the necessary protein, carbohydrates, fats, vitamins, and minerals that chickens need to maintain optimal health and productivity.

One of the most critical components of a chicken's diet is protein. Protein is essential for muscle development, feather growth, and egg production. Commercial poultry feeds typically contain a high percentage of protein derived from sources such as soybean meal, corn gluten meal, fish meal, or meat and bone meal. The protein content of poultry feed varies depending on the chickens' intended use and life stage, with higher protein levels needed for growing chicks and laying hens.

In addition to protein, chickens require carbohydrates for energy to support their daily activities and metabolic functions. Carbohydrates are the primary energy source in a chicken's diet and are found in corn, wheat, barley, and oats. These grains provide chickens the energy they need to maintain body temperature, move around, and perform essential tasks such as foraging and dust bathing.

Fat is another vital component of a chicken's diet, providing concentrated energy and essential fatty acids that support healthy growth, egg production, and immune function. Fat sources commonly found in poultry feed include vegetable oils, animal fats, and seeds such as sunflower or flaxseed. While chickens require fat in their diet, it's essential to provide it in moderation to prevent obesity and related health issues.

In addition to macronutrients like protein, carbohydrates, and fats, chickens also require a variety of vitamins and minerals to maintain overall health and well-being. These include vitamins such as vitamin A, vitamin D, vitamin E, and B vitamins and minerals such as calcium, phosphorus, potassium, and magnesium. Commercial poultry feeds are typically fortified with vitamins and minerals to ensure chickens receive the essential nutrients they need to thrive.

In addition to commercial poultry feed, chickens can benefit from supplemental foods and treats that provide additional nutrition and enrichment. Fresh fruits and vegetables, mealworms, earthworms, and kitchen scraps can be offered as occasional treats to provide variety and stimulation to a chicken's diet. However, it's essential to offer these treats in moderation and avoid feeding foods high in salt, sugar, or fat, as these can be harmful to chickens in excess.

Water is another critical component of a chicken's diet and should be provided in clean, fresh, and easily accessible sources at all times. Chickens require water for hydration, temperature regulation, digestion, and egg production, which is essential for their overall health and well-being. Ensure chickens have access to clean water at all times, especially during hot weather or when laying eggs.

We are surrounded by many nutrients that go unused for human nutrition. Many food co-products, such as dried distiller grains, bakery waste, oilseed meals, and by-products of animal agriculture like meat and bone, blood, or feather meal, all have nutritional value. Still, they are the sort of things that humans won't eat. By raising chickens, we can take what are (to us) inedible ingredients and transform them into lean meat and nutritious eggs.

The birds can use their metabolism to break the nutrients in these feed ingredients into pieces and reassemble them into something we enjoy eating. It's only possible if we provide them with all the nutritional building blocks necessary for the job. Animal feed may be a significant expense in animal production (about 70% of the total cost), but providing your flock with a well-balanced diet and plenty of excellent, clean water means you can receive the maximum supply of fresh eggs or nutritious meat from them.

In conclusion, understanding chickens' dietary needs is essential for chicken keepers to provide their feathered friends with the nutrition they need to thrive. A balanced diet that includes high-quality commercial poultry feed, supplemented with fresh water, occasional treats, and supplemental foods, ensures that chickens receive the essential nutrients, vitamins, and minerals they need for optimal health, productivity, and well-being. Chicken keepers can ensure their flock remains healthy, happy, and productive for years by meeting their dietary needs.

Selecting the Right Feed and Supplements

Selecting suitable feed and supplements is paramount to chickens' health, productivity, and well-being. With many available options, choosing the appropriate feed and supplements can seem daunting for chicken keepers. However, understanding the nutritional needs of chickens and considering factors such as age, breed, purpose, and environmental conditions can help inform your decision-making process and ensure that your flock receives the optimal nutrition they need to thrive.

The foundation of a chicken's diet is a high-quality commercial poultry feed explicitly formulated to meet their nutritional requirements. When selecting poultry feed, it's essential to consider your chickens' life stage and purpose. Starter feeds are formulated for young chicks to support rapid growth and development, while grower feeds are designed for growing chickens to support muscle development and skeletal health. Layer feeds are formulated for laying hens to support egg production and shell quality, while broiler feeds are formulated for meat-producing chickens to support muscle growth and weight gain.

In addition to considering your chickens' life stage and purpose, it's essential to select poultry feed appropriate for their breed and size. Different breeds of chickens have different nutritional requirements, with some breeds being more active or requiring higher levels of certain nutrients than others. Additionally, large breeds such as Brahma or Jersey Giant chickens may require feed with lower calcium levels to prevent skeletal issues. In comparison, smaller breeds, such as bantams, may require feed with higher protein levels to support their metabolic needs.

Different rations are often used, depending on the production stage of the bird. Starter rations are high in protein-an expensive feed ingredient. However, grower and finisher rations can be lower in protein since older birds require less. A starter diet is about 24% protein, a grower diet is 20% protein, and a finisher diet is 18% protein. Layer diets generally have about 16% protein. Special diets are available for broilers, pullets, layers, and breeders. Whole grains can also be provided as scratch grains. Access to clean water is essential. Total dissolved solids above 3000 ppm in the water can interfere with poultry health and production.

Some producers mix their rations to ensure that only "natural" ingredients are used. Poultry feed ingredients include energy concentrates such as corn, oats, wheat, barley, sorghum, and milling by-products. Protein concentrates include soybean meal and other oilseed meals (peanut, sesame, safflower, sunflower, etc.), cottonseed meal, animal protein sources (meat and bone meal, dried whey, fish meal, etc.), grain legumes such as dry beans and field peas, and alfalfa. Grains are usually ground to improve digestibility. Soybeans typically need to be heated-usually by extruding or roasting-before feeding to deactivate a protein inhibitor. Soybeans are generally fed in the form of soybean meal, not in "full fat" form, because the valuable oil is extracted first. Whole roasted soybeans are high in fat, providing birds with energy.

Chicken feed usually contains soybean meal, a by-product of the oilseed industry. In the industry, soybeans are dehulled and cut into thin pieces (flaked) to improve the action of the solvent (usually hexane), which is passed through the soybean to extract the valuable oil. Vegetable oils such as soybean oil are used for edible and industrial purposes. The soybean is then toasted as a method of heat treatment to deactivate an inhibitor that would otherwise interfere with protein digestion in the animal.

However, chickens can also be fed unextracted (full fat) soybeans. An advantage of feeding unextracted soybeans is that they still contain the oil, providing high-energy fat to the bird. Unextracted soybeans must be heat-treated-roasted with dry heat, then ground rolled, or flaked before mixing into a diet. Another method of heat treatment is extruding. Extrusion involves forcing the beans through die holes in an expander-extruder, which creates friction that heats the beans sufficiently (sometimes steam is also applied). The result is a powdery material which does not require further grinding. Roasted and extruded soybeans should only be stored briefly, especially in hot weather, because the oil turns rancid.

Since protein is generally one of the most expensive feed ingredients, the industry uses targeted rations and reduces the amount of protein in the diet as the birds grow (chickens require less and less protein as they age); however, it may not be cost-effective for small-scale producers to have different diets for starters, growers, and finishers.

Furthermore, environmental factors such as climate, temperature, and humidity can also influence chickens' nutritional needs and the type of feed they require. In hot climates, chickens may require feed with lower protein levels to reduce metabolic heat production. In contrast, in cold climates, they may require feed with higher energy levels to maintain body temperature. Additionally, chickens in free-range or pasture-based systems may have access to a more varied diet and require supplemental feed to meet their nutritional needs.

Supplements can play a valuable role in enhancing the nutritional value of a chicken's diet and addressing specific health or performance concerns. Joint supplements for chickens include calcium supplements to support eggshell quality, probiotics to promote digestive health, and omega-3 fatty acids to support immune function and egg production. When selecting supplements for your flock, choosing products appropriate for their age, breed, and nutritional needs is essential and following dosing instructions carefully to avoid over-supplementation.

In addition to commercial poultry feed and supplements, chickens can benefit from occasional treats and supplemental foods that provide variety and enrichment to their diet. Fresh fruits and vegetables, mealworms, earthworms, and kitchen scraps can be offered as occasional treats to provide additional nutrition and stimulation to your flock. However, it's essential to offer these treats in moderation and avoid feeding foods high in salt, sugar, or fat, as these can be harmful to chickens in excess.

When selecting feed and supplements for your flock, choosing products from reputable manufacturers that adhere to high-quality standards and are free from contaminants or harmful additives is essential. Look for products that are labeled as complete and balanced for your chickens' intended life stage and purpose and avoid products that contain unnecessary fillers or artificial ingredients. Additionally, consider consulting with a veterinarian or poultry nutritionist for personalized recommendations based on your flock's specific needs.

In conclusion, selecting suitable feed and supplements is crucial for chickens' health, productivity, and well-being. By considering factors such as age, breed, purpose, and environmental conditions, chicken keepers can make informed decisions that ensure their flock receives the optimal nutrition they need to thrive. Whether choosing commercial poultry feed, supplements, or occasional treats, prioritizing quality and appropriateness ensures that your chickens remain healthy, happy, and productive for years.

Feeding Schedules and Practices

Establishing proper feeding schedules and practices is crucial for chickens' health, productivity, and overall well-being. A well-balanced diet, provided in appropriate quantities and at the correct times, ensures chickens receive the essential nutrients they need to thrive. Moreover, implementing consistent feeding practices helps maintain healthy growth, optimal egg production,

One of the fundamental aspects of feeding chickens is establishing a regular feeding schedule. Chickens thrive on routine and providing them with consistent mealtimes helps regulate their metabolism and digestive processes. Feeding schedules may vary depending on the age, breed, and purpose of your chickens. For example, young chicks may require frequent daily feedings to support their rapid growth and development. At the same time, adult chickens may be fed once or twice daily to maintain their energy levels and overall health. Establishing a feeding routine and sticking to it helps chickens feel secure and confident in their environment, reducing stress and promoting overall well-being.

Moreover, the quantity and quality of feed provided to chickens are essential in feeding practices. Providing chickens with a well-balanced diet that meets their nutritional requirements for protein, carbohydrates, fats, vitamins, and minerals is crucial. Commercial poultry feeds explicitly formulated for different life stages and purposes provide a convenient and reliable source of nutrition for chickens. Additionally, supplementing their diet with fresh fruits, vegetables, and occasional treats can provide variety and enrichment. However, avoiding overfeeding or underfeeding chickens is essential, as both can harm their health and productivity. Monitoring their body condition, egg production, and overall health can help determine if adjustments need to be made to their feeding regimen.

Another essential consideration in feeding practices is always providing access to clean, fresh water. Water is necessary for hydration, temperature regulation, digestion, and egg production in chickens. Ensure chickens have access to clean water sources free from contaminants and debris. In hot weather or during increased activity, chickens may require more water to stay hydrated and maintain their body temperature. Regularly clean and refill waterers to ensure chickens can access fresh water throughout the day.

Regardless of the quality of chicks purchased, good results can only be expected if chicks are fed a nutritious diet. Free choice allows birds to balance or regulate their intake of grain and mash. The free-choice system can work well with small flocks but leaves too much guesswork for a commercial flock. There are, however, general recommendations for feeding replacement chicks, layers, broilers, and turkeys. Controlled mash and grain feeding is used successfully in small flocks. More care and attention must be given to the flock fed this way than the flock fed with all-mash.

Controlled mash and grain feeding involves the use of a concentrate (20 to 23 percent protein), limited amounts of whole grains, and calcium supplements. The amount of grain fed should be calculated to make the total feed intake contain about 16 percent protein. Slight adjustments can be made seasonally to provide more energy in cold weather (more grain) and less energy in hot weather (less grain). Grains may be placed in hoppers or scattered in the litter.

The all-mash system is the most common, especially true for commercial-size flocks. Man takes care of the proper nutritional balance rather than letting the hen balance her own ration. In larger flocks, this ensures against the possibility that the hen will need to balance her ration better. All-mash feeding involves using a single mash containing all components of a balanced ration. Egg quality factors, such as shell thickness and yolk color, are more uniform and more easily controlled with this system. An all-mash ration can be easily dispensed in hanging or automatic feeders. Less skill on the part of the flock owner is required with the all-mash system.

A calcium supplement such as crushed oyster shell or limestone granules can be made a free choice if desired. Also, a portion (1 / 2 or 2/ 3) of the calcium supplement in the ration may be more extensive particle size limestone or oyster shell. Larger calcium supplements take longer to move through the digestive system, so more calcium is absorbed. Small particle-size supplements move quickly through the digestive system, releasing less calcium into the bloodstream.

Additionally, feeding practices should consider the specific needs and requirements of different breeds of chickens. Some breeds may have higher energy requirements or be more prone to particular health issues, such as obesity or egg binding. Tailoring feeding practices to meet the unique needs of your flock can help prevent health problems and promote overall well-being. Consult with a veterinarian or poultry nutritionist for personalized recommendations based on your chickens' specific breed, age, and purpose.

Furthermore, feeding practices should consider environmental factors such as climate, temperature, and seasonal changes. Chickens' nutritional needs may vary depending on environmental conditions, with adjustments needed to ensure they receive the appropriate energy levels, protein, and other nutrients. In hot weather, chickens may require feed with lower protein levels to reduce metabolic heat production. In contrast, they may require feed with higher energy levels to maintain body temperature in cold weather. Additionally, chickens in free-range or pasture-based systems may have access to a more varied diet and require supplemental feed to meet their nutritional needs.

In conclusion, feeding schedules and practices are essential to effective flock management and poultry farming. By establishing regular feeding schedules, providing a well-balanced diet, and considering factors such as breed, age, purpose, and environmental conditions, chicken keepers can ensure their flock receives the nutrition they need to thrive. Moreover, monitoring their body condition, egg production, and overall health allows for adjustments to their feeding regimen as needed. Ultimately, prioritizing proper feeding practices promotes chickens' health, productivity, and overall well-being, ensuring that they remain happy and healthy members of the flock for years to come.

CHAPTER VII

Health and Veterinary Care

Common Health Issues for Chicken

Keeping backyard chickens is a fun and relatively easy experience. However, just like a household pet, chickens can fall ill at times. While the prospect of chickens becoming sick can be scary, the most common health problems in chickens can usually be treated easily from home. Maintaining the health and well-being of chickens is a top priority for chicken keepers, as healthy chickens are more productive, resilient, and enjoyable to care for. However, like all living beings, chickens are susceptible to a variety of health issues that can impact their overall health and quality of life. The key to managing these issues is recognizing the signs of illness. This understanding is essential for chicken keepers to implement appropriate preventive measures and seek timely veterinary care when needed, thereby safeguarding the health of their flock.

One of the most common health issues that chickens face is respiratory diseases. These can be caused by a variety of factors, including viral, bacterial, fungal, and environmental factors such as poor ventilation, overcrowding, and high humidity. However, the risk of these diseases can be significantly reduced through preventive measures such as proper ventilation, biosecurity protocols, and vaccination. By implementing these measures, chicken keepers can play a proactive role in maintaining the health of their flock, ensuring that their chickens remain productive, resilient, and enjoyable to care for.

Another common health issue for chickens is parasitic infestations. Chickens can be affected by various internal and external parasites, including worms, mites, lice, and fleas. Internal parasites such as roundworms, tapeworms, and coccidia can cause weight loss, diarrhea, decreased egg production, and even death if left untreated. External parasites such as mites, lice, and fleas can cause irritation, feather loss, skin lesions, and anemia in chickens. Preventive measures such as regular deworming, parasite monitoring, and maintaining clean, dry living conditions can help prevent parasitic infestations in chickens.

Furthermore, nutritional deficiencies are a common health issue for chickens, particularly if they are not receiving a balanced diet that meets their nutritional needs. This underscores the importance of providing a balanced diet to chickens. Common nutritional deficiencies in chickens include deficiencies in protein, vitamins, and minerals such as calcium, phosphorus, and vitamin D. These deficiencies can lead to a variety of health problems in chickens, including poor growth, weak bones, decreased egg production, and increased susceptibility to diseases. By ensuring that chickens receive a balanced diet that includes high-quality commercial poultry feed, supplemented with fresh water, occasional treats, and appropriate supplements, chicken keepers can help prevent these deficiencies and promote the overall health and well-being of their flock.

Moreover, egg-related issues are common health concerns for laying hens. Egg-related issues such as egg binding, soft-shelled eggs, and egg peritonitis can occur due to genetics, age, diet, and stress. Egg binding occurs when an egg becomes stuck in the oviduct, leading to difficulty laying eggs and potentially fatal complications if left untreated. Soft-shelled eggs occur when hens are deficient in calcium or vitamin D, making thin or fragile eggshells prone to breakage. Egg peritonitis occurs when an egg becomes trapped in the abdomen, leading to inflammation, infection, and potentially fatal complications. Preventive measures such as providing a balanced diet, maintaining proper calcium levels, and monitoring egg production can help reduce the risk of egg-related issues in laying hens.

In addition to these common health issues, chickens can be affected by various other ailments and conditions, including bumblefoot, egg peritonitis, Marek's disease, and Newcastle disease. Bumblefoot is a bacterial infection of the foot that occurs due to injuries or poor husbandry practices, leading to swelling, inflammation, and abscess formation. Egg peritonitis is an inflammatory condition of the reproductive tract that occurs when an egg becomes trapped in the abdomen, leading to infection and potentially fatal complications. Marek's disease is a viral disease that affects the nervous system and immune system of chickens, leading to paralysis, tumors, and increased susceptibility to other diseases. Newcastle disease is a viral disease that affects chickens' respiratory, digestive, and nervous systems, leading to respiratory signs, diarrhea, neurological signs, and high mortality rates. Preventive measures such as biosecurity protocols, vaccination, and proper management practices can help reduce the risk of these and other health issues in chickens.

In conclusion, understanding the common health issues that chickens may encounter is essential for chicken keepers to ensure their flock's health, productivity, and well-being. By recognizing signs of illness, implementing appropriate preventive measures, and seeking timely veterinary care, chicken keepers can help prevent and manage common health issues in chickens. Ultimately, prioritizing proactive health management promotes optimal health and performance in chickens, ensuring they remain healthy, happy, and productive for years.

Preventative Health Care and Vaccinations

Preventative health care and vaccinations are essential components of maintaining chickens' health, well-being, and productivity. As with any animal, chickens are susceptible to various diseases, parasites, and health issues that can impact their overall health and productivity. By implementing preventative health care measures and vaccinating against common diseases, chicken keepers can minimize the risk of illness and promote the longevity and vitality of their flock.

One of the cornerstones of preventative health care for chickens is maintaining a clean and sanitary living environment. Regularly cleaning and disinfecting the coop and surrounding areas helps prevent the buildup of harmful pathogens and parasites that can lead to disease. Removing feces, bedding, and other debris from the coop and regularly replacing bedding materials helps reduce the risk of bacterial and fungal infections. It promotes a healthy living environment for your flock.

In addition to maintaining a clean-living environment, implementing biosecurity measures can help prevent the introduction and spread of diseases within your flock. Biosecurity practices such as limiting exposure to wil birds and other animals, quarantining new birds before introducing them to your existing flock, and restricting access to visitors and equipment from other poultry farms can help minimize the risk of disease transmission and protect the health of your chickens.

Another important aspect of preventative health care for chickens is providing a balanced diet that meets their nutritional needs. Proper nutrition is essential for maintaining a robust immune system, promoting growth and development, and supporting egg production in laying hens. Offering high-quality commercial poultry feed formulated specifically for your chickens' life stage and purpose, as well as supplementing their diet with fresh fruits, vegetables, and occasional treats, ensures that your flock receives the essential nutrients they need to thrive.

Vaccinations are critical tools in preventing the spread of infectious diseases and protecting chickens' health. Vaccines are available for common poultry diseases, including Marek's disease, Newcastle disease, infectious bronchitis, and avian influenza. Vaccinating against these diseases helps reduce the risk of illness, mortality, and economic losses associated with outbreaks, particularly in commercial poultry operations.

The timing and frequency of chicken vaccinations depend on factors such as the prevalence of specific diseases in your area, the age and breed of your chickens, and your management practices. Some vaccines are administered to day-old chicks through the incubator, while others require multiple doses at specific intervals to provide long-lasting immunity. Working with a veterinarian or poultry health professional can help you develop a vaccination program tailored to the needs of your flock and ensure that your chickens receive the appropriate protection against common diseases.

In addition to vaccinations, regular health checks and monitoring are essential for identifying and addressing potential health issues in your flock. Observing your chickens' behavior, appetite, egg production, and overall appearance can provide valuable insights into their health status and help you detect signs of illness or injury early. Regularly examining your chickens for signs of parasites such as mites, lice, and worms and treating them promptly with appropriate medications helps prevent infestations and keeps your flock healthy and comfortable.

Furthermore, providing access to fresh water is essential for maintaining hydration, promoting digestion, and supporting overall health and well-being in chickens. Water is involved in virtually every aspect of a chicken's physiology, from temperature regulation and nutrient transport to waste elimination and egg production. Ensuring your chickens have access to clean, uncontaminated water always helps prevent dehydration, heat stress, and other health issues associated with inadequate hydration.

In conclusion, preventative health care and vaccinations are essential for maintaining chickens' health, productivity, and well-being. By implementing biosecurity measures, maintaining a clean-living environment, providing a balanced diet, and vaccinating against common diseases, chicken keepers can minimize the risk of illness and promote the longevity and vitality of their flock. Whether raising chickens for eggs, meat, or companionship, prioritizing preventative health care measures ensures your flock remains healthy, happy, and productive for years.

Recognizing and Treating Illnesses

Recognizing and treating illnesses in chickens is a critical skill for any poultry keeper. Like all animals, chickens are susceptible to a variety of diseases, infections, and health issues that can impact their overall well-being and productivity. By learning to recognize the signs and symptoms of common chicken illnesses and implementing appropriate treatment strategies, chicken keepers can minimize the impact of illness on their flock and promote the health and longevity of their birds.

Chicken infections must be avoided whether you are growing hens in your garden or for commercial purposes. Chickens are very simple to care for, but things may quickly become problematic when they become ill. Many poultry farms fail because they cannot control specific chicken diseases that strike their flock. I'm sure you want to avoid winding up like one of those farms, which is why you're reading this.

Chicken illnesses are simply diseases that harm chickens. These infections may make your hens very sick and perhaps kill them. Some chicken diseases have the potential to transmit to people, resulting in catastrophic illnesses. Salmonellosis, campylobacteriosis, and avian influenza viruses are among them. As a poultry farmer, you must do all possible to keep these deadly illnesses from killing your hens. Types of chicken illnesses are classified based on how they impact the birds. There are several poultry illnesses out there ready to infect your animals. As a result, it is critical to understand how they join the flock so that you can readily recognize and eradicate them.

This sort of chicken sickness is widespread and readily transmitted on your farm from one bird to the next. Viruses, bacteria, and fungi are the most common causes of infectious poultry illnesses. When viruses, bacteria, and fungi infiltrate your poultry farm, they harm the birds' intestines, nerves, and lungs. They also hurt the bird's immunological and reproductive systems, as well as its skin. Because infectious illnesses spread so rapidly on farms, isolating any bird that exhibits symptoms of the disease is preferable.

One of the first steps in recognizing and treating illnesses in chickens is becoming familiar with the signs and symptoms of common diseases and health issues. Common signs of illness in chickens include lethargy, decreased appetite, abnormal droppings, respiratory distress, swollen eyes or sinuses, changes in behavior or posture, and reduced egg production in laying hens. Additionally, chickens may exhibit specific symptoms depending on the type of illness they are experiencing, such as coughing, sneezing, wheezing, or lameness.

Once the signs of illness are recognized, prompt action is essential to address the problem and prevent it from spreading to other members of the flock. Isolating sick chickens from the rest of the flock and providing them with a separate, comfortable living space can help prevent the spread of illness and facilitate their recovery. Additionally, seeking veterinary advice and treatment as soon as possible can help diagnose the underlying cause of the disease and determine the most appropriate course of action.

Treatment strategies for chicken illnesses vary depending on the disease or health issue. In some cases, supportive care, such as providing warmth, fluids, and nutritional support, may be sufficient to help chickens recover from illness. In other cases, medications such as antibiotics, antivirals, or antiparasitic may be necessary to treat bacterial infections, viral diseases, or parasitic infestations. It's essential to follow veterinary recommendations and treatment protocols carefully to ensure the safety and efficacy of any chicken medications.

In addition to conventional veterinary treatments, some chicken keepers may explore alternative or complementary therapies to support the health and well-being of their flock. Herbal remedies, homeopathic treatments, and dietary supplements are examples of alternative therapies that some chicken keepers may use to address specific health issues or promote overall wellness in their birds. While these treatments may have anecdotal or historical support, it's essential to use caution and consult a veterinarian before implementing alternative or complementary therapies for chicken illnesses.

Preventing illness in chickens is always preferable to treating it after it occurs. Implementing biosecurity measures, maintaining a clean and sanitary living environment, and providing a balanced diet with access to fresh water are essential to preventing illness and promoting overall health in chickens. Additionally, practicing good hygiene, such as washing hands before and after handling chickens, wearing dedicated footwear in the coop area, and regularly cleaning and disinfecting equipment and supplies, helps minimize the risk of disease transmission between chickens and humans.

Regular health checks and monitoring are also crucial for identifying and addressing potential health issues in chickens before they escalate into more severe problems. Observing chickens' behavior, appetite, egg production, and overall appearance daily can help detect signs of illness or injury early and facilitate prompt intervention and treatment. Additionally, regularly examining chickens for external parasites such as mites, lice, and ticks and treating them promptly with appropriate medications helps prevent infestations and keeps chickens healthy and comfortable.

In conclusion, recognizing and treating illnesses in chickens is an essential skill for poultry keepers to ensure their flock's health, well-being, and productivity. By learning to recognize the signs and symptoms of common chicken illnesses, implementing appropriate treatment strategies, and practicing good biosecurity and hygiene measures, chicken keepers can minimize the impact of disease on their flock and promote a healthy and thriving chicken community.

CHAPTER VIII

Understanding Chicken Behavior

Common Chicken Behaviors and What They Mean

Understanding chicken behavior makes management more accessible and efficient and helps you provide your chickens with the best quality of life. Maintenance behaviors help sustain physiological equilibrium. These include feeding and foraging, drinking, resting, and comfort. Chickens spend around 61 percent of their active time on foraging and feeding behavior. Foraging behaviors include pecking and scratching at the ground to find food sources. Feeding behaviors involve consuming and ingesting food.

Foraging is a highly motivated behavior that chickens perform even when unnecessary. That is called contra-freeloading, which means the chickens work for food even when readily available. You may notice that your chickens are eating and foraging simultaneously. It's because both feeding and foraging are social behaviors.
If a chicken sees another chicken feeding, it also wants to start feeding. This helps them find food and lowers the predation risk. Synchronized foraging means the birds take turns looking out for predators while the rest eat, instead of each having to forage and be vigilant. If a chicken cannot perform foraging behaviors, it can become frustrated and exhibit unwanted abnormal behaviors, such as aggressive feather pecking, egg eating, and cannibalism.

Resting and sleeping are essential to chickens, preparing them for activities of the upcoming day. While sleeping, all the bird's memories are consolidated and stored.

Although chickens can rest on the ground, they prefer to rest on perches. The behavior of resting on a perch is called roosting. Roosting allows the birds to sleep elevated and protected from any ground predators. Chickens usually start roosting around dusk. Perches also provide somewhere for subordinate birds to escape the harassment of more dominant birds. The ability to roost and use perches improves the bird's bone strength, foot health, and feather condition. Chicks start roosting at 1 to 2 weeks of age.

Chickens perform behaviors related to body care and maintenance. These are called comfort behaviors and involve taking care of plumage and stretching. Examples include dust bathing, preening, leg and wing stretching, flapping, and tail wagging. Preening is the chicken's version of grooming, ensuring its feathers are in good condition. When preening, a chicken runs its beak through its feathers, realigning the barbs and barbules. Allows the feathers to function correctly. The chicken removes any debris or external parasites within its plumage during preening.

Another part of preening is oiling the feathers. In this process, the bird takes oil from it preen gland into its beak and distributes it along the feathers. Preening and the distribution of preen oil help keep feathers healthy and provide insulation and waterproofing. Preening is a social behavior. Chickens are often seen synchronously preening as a larger group rather than an individual. Comfort behavior that chickens perform is dust bathing. Rather than bathing in water, chickens bathe in dust. When dust bathing, the chicken digs a small hollow, then lays down and rolls around in the dirt (or other substrate), rubbing it into its feathers. It then stands up and shakes all the substrate from its feathers.

Although the type of substrate is essential in how effective it is in removing external parasites, the performance of dust bathing itself helps chickens get rid of dead skin and stale preen oil. Dust bathing helps the chicken with insulation, waterproofing, and maintaining the down condition. Just like preening, dust bathing is a synchronous social behavior. Both dust bathing and preening are highly motivated behaviors. If chickens are prevented from performing these behaviors, frustration and behavioral issues can ensue. These can include pacing back and forth, sham dust bathing (imitation dust bathing when no substrate is present), repetitive pecking at one spot, pecking and pulling at feathers from other birds in the flock, and making calls. Addressing any stress or frustration in the flock is essential, as this can cause birds to peck at each other and may lead to cannibalism.

Chickens spend a lot of time exploring. This behavior starts on the first day of life when the chick pecks at potential food objects. Exploratory behavior consists of pecking at objects and the environment with the beak and scratching the floor. The main chicken exploratory organ is the beak. Its tip contains many nerve endings with touch receptors that provide sensory information to the bird. The bird learns about it by pecking at an object and the environment. Another important part of chicken behavior is social behavior, which involves interaction with other chickens.

Chickens usually live in small groups with a very definite social hierarchy. Within flocks is a clearly defined ranking of which birds are the dominant birds (high in the hierarchy) and which are subordinate (bottom of the hierarchy). Roosters and hens have their separate hierarchy within a flock. The pecking order starts to establish itself about one week after hatching and is fully established around six weeks of age. Once this pecking order is established and stable, the birds will happily live together unless a new bird is added, or an unusual incident occurs.

Chickens recognize the other chickens within the flock, know their standing within the pecking order, and know whether they are dominant or subordinate to them. A subordinate bird performs certain submissive behaviors within the pecking order to show its submission to the dominant birds. These behaviors include escaping (running away) and crouching or squatting. If a hen squats or crouches to the flock owner, she is demonstrating her submission to the owner. Chickens are social learners, meaning that they can learn a new behavior by observing a flock mate perform it. They can also learn from observing humans.

If a new chicken comes into a flock, current flock residents can observe where the newcomer's place is within the pecking order and where they rank about the newcomer's status. A hen will watch an interaction between a new chicken and a dominant flock mate (one higher in the pecking order). If the dominant flock mate loses to the newcomer, then the observing hen knows that the new bird is higher in the pecking order and will not challenge it. However, if the new bird loses, the observing bird will challenge the newcomer to discover their relationship.

Chickens also use social learning when they are chicks. By observing their mothers, chicks learn what is good to eat, where to find food, how to perch, and where their home range is. Chicks even learn by observing other chicks. If they watch another chick eat something and react in disgust, the observing chicks learn not to try that food. Chickens communicate primarily through displays of body language and vocalizations. Displays are often used to communicate if flock mates are within an intermediate distance from them. Vocalizations are used to communicate with flock mates and birds from other flocks farther away. Displays involve changing the posture and position of the head and body. These changes could include holding the head at an angle, raising or lowering the tail, and spreading the feathers wide or flat. These are all essential signals about personal space, health, and group organization; they are also displayed during mating and territory disputes.

Vocalizations are the different sounds or calls that chickens use to communicate with each other. It is their version of talking. These vocalizations can communicate with chickens within the same flock or in different flocks. For example, the rooster crow call can be used to defend his territory from other roosters without fighting. Chickens perform over thirty vocalizations, each communicating a different type of information, such as contentment, pleasure, frustration, distress, fear, danger, nesting, food, courtship, and territory. When a chicken hears one of these vocalizations, it understands the information and responds accordingly. If a chick makes a distress call, the mother hen immediately looks for it. Similarly, if a hen clucks at her chicks, she calls them to come to her. Her chicks will respond by approaching her and gathering under her wings. Chickens have two different alarm calls —one for aerial predators and another for ground predators. The behavioral responses are different for each call.

Chickens use vocalization when they are still in the shell to communicate with the mother before they hatch. Chicks also communicate with chicks in other eggs, getting them to accelerate their growth to hatch around the same time. Some behaviors that chicks perform are instinctive, while others must be learned. Preening, scratching at the ground, and responding to the mother hen are all instinctive behaviors. Behaviors such as drinking are not instinctual and need to be taught.

Chicks learn by imprinting: they instinctively follow and learn from the first moving object. Usually, this is the mother hen, and she will teach them all the essential behaviors, such as drinking, foraging, the home range, perching, and social relationships. If no mother is present, the chicks will imprint on the other chicks. Another chick-specific behavior is play. While performing play behavior, chicks frolic and spar together. These little play fights are practiced for the adult fights that occur later in life when establishing a pecking order.

In Conclusion, chickens have a wide range of behaviors they perform. Maintenance behaviors, such as foraging, drinking, resting, preening, and dust bathing, help maintain physiological equilibrium. Social behaviors involve interaction with one another. Aspects of this are pecking order, how they behave in a social group, learn from each other, and communicate via displays and vocalizations. Chicks have some behaviors that are specific to them, such as play behavior and a particular way of learning, namely imprinting. Overall, it is essential to understand a chicken's behavior so that you can give it the best quality of life possible.

Social Dynamics and Pecking Order

Understanding the social dynamics and pecking order within a chicken flock is fundamental for any poultry keeper aiming to maintain a healthy and harmonious group of birds. Chickens are inherently social animals that establish complex social hierarchies, which dictate access to resources and influence overall flock behavior. By comprehending these social structures and behaviors, chicken keepers can better manage their flocks, reduce conflict, and ensure the well-being of their birds.

Chickens establish their pecking order out of instinct. They use this hierarchy to determine the order in which they eat and drink. The pecking order also affects activities like roosting, egg-laying, and mating. A chicken's life is influenced by its place in the pecking order. More robust, dominant chickens at the top of the order first get access to food, water, and treats. They also get their pick of roosting space. Chickens at the lower end of the pecking order must wait their turn for these things.

When lower-ranked chickens try to usurp their flock members by getting to these things first, they'll get knocked back into place with a peck or other show of dominance. Of course, these rankings are only sometimes permanent. As chickens age and new members join the flock, everyone's place in the pecking order can shift. Some dominant chickens will even relinquish their spot as they grow older or tired of their duties. The chicken at the top of the pecking order has plenty of responsibilities to go alongside its privileges. This chicken is the strongest and healthiest of the flock. As such, they play the role of flock protector. They practice constant vigilance, watching for predators and other dangers to their little feathery family.

If a hawk flies overhead, the chicken at the top of the pecking order is the one to herd the flock to the safety of the coop. If there are disputes between other flock members, the alpha chicken will step in to settle it. Some flock mates will even stand back and let the rest of the flock eat first while they keep watch. If you have a rooster in your established flock, he will most likely take the top chicken spot of the pecking order. If you have other roosters in the flock, they'll take their natural places throughout the hierarchy.

With no roosters in a flock, an older, more assertive, and dominant hen will be alpha for flock management. That said, young pullets tend to be less violent with their shows of dominance, leading to a calmer establishment of the pecking order. As we mentioned above, chickens establish and maintain their roles in the pecking order through shows of dominance. These incidents are often quick and mostly harmless.

If a chicken wants to confront another flock member, it might strut about, flap its wings, fluff its feathers, and squawk at the other young birds. Sometimes, that's all it takes, and the other flock member will concede, establishing that the challenger is higher than they are in the pecking order. However, if the second chicken doesn't concede, the confrontation can escalate to pecking and squabbling. It might also end quickly with nothing more than a few lost feathers. However, if neither chicken backs down, they will keep fighting. In extreme situations, this can lead to severe injury or even death.

While establishing the pecking order is natural, you want to watch your flock. Once a dispute escalates to drawing blood, it's time to break it up and separate the birds for a while. Introducing new members to the flock will mix up the pecking order. It may take time for your latest chickens to find their place in the hierarchy. It can be stressful for both you and your flock.

Make the process easier by introducing the new birds slowly. You can section off a portion of the coop or run to hold your new chickens for a week. That will allow the birds to get to know each other without physically sharing space, preventing immediate and violent squabbling. Once the initial wariness and aggravation fade, try to let the new birds out to meet the rest of the flock. Squabbling will likely occur, but your chickens will usually settle their issues quickly and develop a new pecking order everyone can live with.

Part of understanding chickens' social dynamics and the pecking order is understanding your place in it. Many chickens see you as a weird, tall, featherless flock member, so you must ensure your chickens know you're at the top. If a particularly aggressive head rooster or hen tries to challenge you, let them know you're the top bird. Never run from an attacking chicken. Instead, try to grab the chicken and hold it gently but firmly to the ground.

Once it calms down, you can let it up again. Repeat this process as needed, and after a couple of incidents, the chicken should respect your place as master of the flock. However, some chickens—usually roosters—continue to be aggressive toward their humans. An overly aggressive bird can be both annoying and dangerous, especially if you have younger children running around.

If you can't correct the behavior, consider removing the chicken by giving it to a different home or using your chicken processing equipment. If you rehome an aggressive chicken, ensure the future owner is aware of their behavior and ready to handle the challenge. The pecking order is natural; chickens usually work together without actual harm. A particularly aggressive chicken or a constantly squabbling flock can be a severe problem. Constant quarrels also put your chickens in danger. Keep everyone safe by giving them a home that doesn't require as much fighting.

When chickens have plenty of access to food, water, roosting spaces, and other necessities, they don't have to argue over who gets access to these resources. Lighting can also stress out your flock. If you have lights in the coop, ensure they are light enough and limit them to only 16 hours a day or less. It will help keep your chickens calm and happy. When you can prevent serious disputes, the pecking order can be fascinating, enlightening, and entertaining.

"Pecking order" describes the hierarchical structure observed in chicken flocks. This hierarchy is established through various social interactions and behaviors, most notably pecking, which determines each bird's rank within the group. The top-ranking chicken, often called the "alpha" or "top hen," enjoys priority access to food, water, nesting sites, and preferred roosting spots. Subsequent ranks follow a descending order of dominance, with the lowest-ranking bird having the least access to these resources.

Establishing the pecking order begins early in a chicken's life, often among chicks within a brood. Chicks quickly learn their place in the social hierarchy through pecking, chasing, and other assertive behaviors. These early interactions can be vigorous but are generally a natural part of flock dynamics. As chickens grow, the pecking order becomes more defined and stable, though it can be disrupted by changes such as the introduction of new birds, illness, or the removal of individuals.

The dominant chickens in the hierarchy, typically the older or more assertive birds, maintain their status through displays of aggression and dominance. These behaviors include pecking at lower-ranking birds, chasing them away from resources, and asserting control over nesting sites and roosting positions. While these interactions may seem harsh, they are a natural part of establishing and maintaining social order within the flock. However, excessive aggression or bullying can indicate issues such as overcrowding, lack of resources, or stress within the flock, which need to be addressed by the keeper.

Understanding a flock's social dynamics is crucial for effectively managing these behaviors. For instance, providing sufficient space, multiple feeding and watering stations, and ample nesting sites can help reduce competition and minimize aggressive interactions. Additionally, ensuring new birds are introduced gradually and carefully can help prevent significant disruptions to the established pecking order. It often involves quarantining new birds, allowing them to adapt to their new environment, and then integrating them slowly with the existing flock under close supervision.

The social structure within a flock is not only about dominance and access to resources but also about social bonds and cooperation. Chickens often form close bonds with specific flock mates, engaging in mutual preening, foraging together, and even roosting close to one another. These social bonds can provide comfort and reduce stress, contributing to the overall well-being of the flock.

Observing these interactions can provide valuable insights into the social structure and relationships within the flock. Behavioral issues such as feather pecking, and cannibalism can arise within the context of social dynamics and the pecking order. Feather pecking involves birds pecking at and pulling out the feathers of their flock mates, which can lead to injury, stress, and increased susceptibility to infections. In severe cases, this behavior can escalate to cannibalism, where birds may peck at wounds and cause significant harm. These behaviors are often a result of stress, boredom, overcrowding, or nutritional deficiencies. Addressing these issues involves improving living conditions, providing adequate space, enriching the environment with perches and foraging opportunities, and ensuring a balanced diet.

The social hierarchy within a flock can also impact egg production and overall productivity. Stressful social interactions and aggressive behaviors can reduce egg production, as hens may become too stressed to lay eggs regularly. Ensuring that hens have access to private, comfortable nesting sites and minimizing stress through proper flock management can help maintain consistent egg production.

Introducing new birds to an established flock requires careful planning and consideration of the existing social dynamics. New birds should be quarantined initially to prevent disease spread and allow them time to acclimate to their new surroundings. Once they are ready to be introduced, it is advisable to do so gradually by placing them in a separate but adjacent pen where they can see and interact with the existing flock without direct physical contact. It allows the birds to become familiar with each other and helps reduce the likelihood of aggressive behavior when they are finally integrated.

Changes in the social hierarchy can occur naturally over time due to various factors such as age, health, and changes in the flock composition. Older birds may lose their dominant status as they age or become less healthy, allowing younger or more assertive birds to move up in the hierarchy. Similarly, removing or adding birds can cause temporary disruptions as the flock reestablishes its social order. Understanding and anticipating these changes can help chicken keepers manage their flocks more effectively and maintain a stable and harmonious group.

In conclusion, the social dynamics and pecking order within a chicken flock are complex and integral aspects of their behavior. By understanding how these hierarchies are established and maintained, chicken keepers can create an environment that supports the natural social structure of their birds while minimizing stress and aggression. Providing adequate space, resources, and enrichment and careful management of flock changes can help ensure a healthy and productive flock. Recognizing the importance of social bonds and cooperation among chickens further enhances their well-being and contributes to a thriving and harmonious flock.

Building Trust and Bonding with Your Chickens

Building trust and bonding with your chickens is a gratifying aspect of poultry keeping that enhances both the keeper's and the flock's experience. Like many animals, chickens can form strong bonds with their caregivers, leading to a more harmonious and enjoyable environment. Establishing trust and building a bond with your chickens requires patience, consistency, and understanding of their behavior and needs.

Building trust with chickens begins with providing a safe and comfortable environment. Chickens are prey animals and naturally cautious, so creating a secure living space where they feel protected from predators and stressors is crucial. A well-designed coop and run, free from potential threats, allows chickens to feel safe and more likely to approach their caregiver without fear. Additionally, ensuring they have access to clean water, nutritious food, and appropriate shelter helps meet their basic needs, fostering a sense of well-being and trust.

Consistent and gentle interaction is essential in building a bond with your chickens. Spending time with them daily, speaking softly, and moving slowly around them helps chickens become accustomed to your presence. Initially, they may be wary or nervous, but regular interaction helps them recognize you as a non-threatening part of their environment over time. Offering treats by hand is an effective way to build trust. Chickens quickly learn to associate their caregiver with positive experiences, and treats can be a powerful motivator. You can encourage chickens to approach and eventually eat from your hand by using small, healthy treats like mealworms, fruits, or vegetables.

Understanding chicken behavior and body language is also essential in fostering trust. Chickens communicate through vocalizations and physical behaviors, indicating their comfort levels and emotional states. For example, softly clucking signifies contentment, while loud squawking may indicate alarm or distress. Observing and responding appropriately to these cues shows chickens you respect their space and are attuned to their needs. Recognizing when a chicken is frightened or stressed and giving them space can prevent negative associations with your presence.

Patience is a virtue when it comes to bonding with chickens. Each bird has its personality and may take a long time to warm up to human interaction. Some chickens are naturally more curious and outgoing, while others may be shy and reserved. Respecting these individual differences and allowing chickens to approach you at their own pace is crucial. Forcing interaction can cause stress and hinder the bonding process. Instead, creating a routine where chickens can gradually become comfortable with your presence and build trust over time is more effective.

Physical contact, such as gentle petting and handling, can strengthen the bond between you and your chickens. Once chickens are comfortable approaching you, gently stroking their feathers or picking them up can further build trust. It is vital to handle chickens correctly to avoid causing them stress or discomfort. Supporting their bodies and avoiding sudden movements helps chickens feel secure while being held. Regular, gentle handling can make chickens more docile and easier to manage, especially during health checks or when moving them to different areas.

Training and enrichment activities can also enhance the bond between you and your chickens. Chickens are intelligent animals capable of learning simple tricks and responding to cues. Training chickens to come when called, perch on your arm, or navigate obstacle courses can provide mental stimulation and strengthen your relationship. Positive reinforcement, which involves giving food and praise, encourages hens to participate in these activities and see their caregiver as a pleasant and rewarding person. By giving hens access to enrichment activities like perches, dust baths, and foraging opportunities, you can keep them happy and involved while strengthening their good associations with their surroundings and caretakers.

Building trust with chickens is about the physical aspects of care and creating a positive emotional connection.

Observing your chickens, talking to them, and being present in their daily lives helps create a bond beyond mere routine. This emotional connection benefits the chickens and the caregiver, fostering a sense of companionship and mutual understanding. Chickens who trust their caregiver are more likely to approach, interact, and seek attention, leading to a more fulfilling and enjoyable experience for both parties.

One of the most rewarding aspects of building trust with chickens is witnessing their unique personalities and behaviors. Chickens are fascinating creatures with distinct characters, and their individuality becomes more apparent as trust builds. Some may be bold and adventurous, while others are gentle and affectionate. This diversity makes each interaction unique and contributes to the overall joy of poultry keeping.

It is also essential to recognize that building trust is an ongoing process. Changes in the environment, health issues, or disruptions in routine can affect chickens' behavior and trust levels. Consistently providing care, attention, and positive interactions helps maintain and strengthen the bond. Being attuned to the needs and behaviors of your chickens ensures that any issues are promptly addressed, maintaining a stable and trusting relationship.

In conclusion, building trust and bonding with your chickens is a multifaceted process involving providing a secure environment, consistent and gentle interaction, understanding their behavior, and respecting their personalities. By developing a strong bond with your chickens, you enhance their well-being and enrich your experience as a poultry keeper. The rewards of this relationship are evident in the trust and companionship that develop, leading to a more harmonious and fulfilling experience for both chickens and their caregivers. As you spend time with your flock, you will discover the deep satisfaction of earning their trust and enjoying their unique companionship.

CHAPTER IX

Egg Production and Care

Understanding the Egg-Laying Process

Understanding the egg-laying process in chickens is essential for anyone involved in poultry keeping, whether for commercial production or backyard farming. The egg-laying process is a complex physiological phenomenon that involves various stages and is influenced by numerous factors, including genetics, nutrition, environment, and health. This section delves into the intricacies of the egg-laying process, providing a comprehensive understanding of how chickens produce eggs and what can be done to support and enhance this natural process.

The egg-laying process begins with the development of the egg within the hen's reproductive system, specifically in the ovary and oviduct. Hens are born with a finite number of ova, or yolk precursors, present in their ovaries from birth. These ova develop into mature yolks through a process called follicular development. When a yolk reaches maturity, it is released from the ovary in a process known as ovulation is the first step in egg formation.

Once the yolk is released, it enters the oviduct, a specialized tube where the rest of the egg formation occurs. The oviduct is divided into several sections, each playing a crucial role in the development of the egg. The first section is the infundibulum, where the yolk is captured after ovulation. Fertilization, if it occurs, happens here. The yolk then moves to the magnum, where it spends approximately three hours. In the magnum, egg white or albumen layers are deposited around the yolk. The albumen provides the necessary proteins and protects the yolk during its development.

Following the magnum, the yolk and albumen move to the isthmus. Here, the inner and outer shell membranes form around the egg, which takes about 75 minutes. These membranes provide structural support and act as barriers to pathogens. The next stage occurs in the shell gland, also known as the uterus, where the egg spends the most prolonged period, about 20 hours. During this time, the eggshell forms. The shell is composed primarily of calcium carbonate deposited in layers. The pigmentation of the shell, which varies among breeds, also occurs in this stage.

The final stage in the oviduct is the vagina, where a protective coating called the cuticle is added to the eggshell. The cuticle helps protect the egg from bacterial contamination and moisture loss. Once the egg is complete, it is laid through the cloaca, the standard exit for the reproductive and digestive tracts. The egg-laying process takes about 24 to 26 hours from ovulation to laying.

Several factors influence the efficiency and consistency of the egg-laying process. One of the most critical factors is nutrition. Hens require a balanced diet of proteins, vitamins, minerals, and essential fatty acids to support egg production. Calcium is necessary for shell formation, and a deficiency can lead to weak or malformed shells. A consistent supply of oyster shells or other calcium supplements can help ensure hens have the necessary nutrients for robust egg production.

Lighting also plays a significant role in regulating egg-laying. Chickens are photoperiodic animals, meaning the length of daylight influences their reproductive cycles. Hens typically require about 14 to 16 hours of light daily to maintain optimal egg production. In commercial settings or during shorter winter days, artificial lighting is often used to simulate longer daylight hours and sustain egg production.

Stress is another factor that can significantly impact egg-laying. Chickens are sensitive to changes in their environment, and stressors such as predation threats, overcrowding, poor ventilation, or abrupt changes in their routine can decrease egg production. A stable, comfortable, and secure environment is crucial for consistent egg production. It includes providing adequate space, clean living conditions, and minimizing disruptions. Health is paramount in the egg-laying process. Diseases, parasites, and other health issues can adversely affect egg production. Regular health checks, vaccinations, and parasite control measures are essential to keep the flock healthy. Additionally, maintaining good biosecurity practices helps prevent the introduction and spread of diseases within the flock.

Understanding the different stages of a hen's life cycle is also essential in managing egg production. Hens reach sexual maturity and begin laying eggs at around 18 to 20 weeks. Peak production usually occurs between 25 to 35 weeks of age, after which egg production gradually declines. When hens reach two to three years, their egg production significantly decreases, although they may still lay sporadically. Knowing this life cycle helps poultry keepers plan for replacement hens to ensure a continuous supply of eggs.

Egg abnormalities can occur and understanding their causes can help in addressing them. Common abnormalities include soft-shelled eggs, which may result from insufficient calcium or other nutritional deficiencies. Double-yolked eggs occur when two yolks are released and encased in one shell, a phenomenon more common in younger hens whose reproductive systems are still maturing. Blood spots and meat spots within eggs are usually harmless and result from minor ruptures in the oviduct or tissue fragments.

Broodiness is a natural behavior in some hens, where they develop the instinct to sit on and incubate eggs. While this behavior benefits natural reproduction, it can interrupt regular egg production in a non-breeding setting. Managing broodiness involves removing the broody hen from the nest, providing a different environment, and sometimes using cooling methods to lower her body temperature and reduce the broody behavior.

Collecting and handling eggs is essential to maintain their quality and safety. Eggs should be collected at least once daily, more frequently in hot weather, to prevent spoilage and reduce the risk of contamination. Gentle handling and proper storage, typically at about 50-60 degrees Fahrenheit with moderate humidity, help preserve egg quality. Washing eggs is a topic of debate; while it can remove dirt and bacteria, it also removes the protective cuticle. If washing is necessary, it should be done with warm water, and the eggs should be dried and refrigerated promptly.

Marketing and utilizing eggs is another crucial aspect of understanding the egg-laying process, particularly for those involved in commercial production. Eggs can be sold fresh, used in various culinary applications, or processed into products like liquid eggs or dried egg powder. Understanding consumer preferences, market demands, and regulations is crucial for successful egg marketing. Proper labeling, including information about production method (e.g., free-range, organic), helps inform consumers and can add value to the product.

Eggs have various by-products that can be utilized in addition to traditional uses. Eggshells, for instance, are rich in calcium and can be ground into powder to supplement animal feed or used as a soil amendment in gardening. Egg membranes contain valuable proteins and are used in cosmetic and biomedical applications. Exploring these alternative uses can add value and sustainability to egg production.

In conclusion, understanding the egg-laying process involves a comprehensive knowledge of the physiological, environmental, and managerial factors that influence egg production. From the intricate process of egg formation within the hen's reproductive system to the factors that impact laying efficiency and egg quality, each aspect plays a crucial role in successful poultry keeping. Poultry keepers can support and enhance the natural egg-laying process by providing proper nutrition, maintaining optimal environmental conditions, managing stress, ensuring health, and understanding the life cycle of hens. Whether for commercial production or backyard farming, this knowledge is essential for providing a healthy, productive, and sustainable flock. The rewards of understanding and managing the egg-laying process extend beyond mere egg production, contributing to the well-being and satisfaction of both the hens and their caregivers.

Enhancing Egg Production

Enhancing egg production is a multifaceted endeavor that requires a thorough understanding of the biological processes involved, the environmental factors that can be manipulated, and the management practices that support optimal production. Whether for commercial purposes or backyard farming, improving egg production involves a combination of genetics, nutrition, housing, health management, and environmental control. This section explores these aspects in detail, providing a comprehensive guide to strategies for maximizing egg production.

Genetics plays a fundamental role in determining a hen's egg-laying potential. Different breeds of chickens have varying capacities for egg production, influenced by their genetic makeup. For instance, commercial hybrid breeds such as the White Leghorn are renowned for high egg production, often laying over 300 eggs annually. These breeds have been selectively bred over generations to enhance their laying performance, feed efficiency, and egg quality. In contrast, heritage breeds like the Rhode Island Red or Plymouth Rock, while still productive, typically lay fewer eggs per year but may offer other advantages such as hardiness, dual-purpose utility, or suitability for free-range conditions. Choosing the right breed or hybrid to match the specific goals and conditions of the poultry operation is a critical first step in enhancing egg production.

Nutrition is one of the most critical factors influencing egg production. Hens require a balanced diet that meets their energy, protein, vitamin, and mineral needs. The primary components of a hen's diet include carbohydrates, proteins, fats, vitamins, and minerals, each playing a unique role in supporting overall health and egg production. Carbohydrates and fats provide the energy necessary for daily activities and egg-laying. Proteins, specifically the amino acids lysine and methionine, are essential for forming egg proteins, including the albumen (egg white) and yolk. Vitamins, particularly A, D3, E, and the B complex are crucial for metabolic processes and overall health. Minerals such as calcium, phosphorus, and magnesium are vital for eggshell formation and structural integrity. Calcium, in particular, is paramount for eggshell quality; hens require a steady supply of dietary calcium, which can be supplemented through sources like oyster shells or limestone. Ensuring a consistent and high-quality feed supply that meets these nutritional requirements is essential for maintaining peak egg production.

Environmental conditions significantly impact egg production. Light exposure, temperature, and housing conditions must be optimized for laying hens. Light is a primary ecological cue for egg production, as chickens are photoperiodic animals. It means their reproductive cycles are influenced by the length of daylight they receive. Hens typically require about 14 to 16 hours of light daily to maintain optimal egg production. During shorter days, particularly in winter, artificial lighting can extend the perceived day length and stimulate egg laying. Properly designed lighting schedules should gradually increase and decrease light exposure to mimic natural patterns and avoid stress.

Temperature also affects egg production, with extreme temperatures reducing laying rates. The optimal temperature range for laying hens is between 55 and 75 degrees Fahrenheit. High temperatures can cause heat stress, reducing feed intake and decreasing egg production. Conversely, very low temperatures can increase hens' energy requirements, diverting resources away from egg production to maintain body heat. Providing adequate ventilation, insulation, and, if necessary, heating or cooling systems can help maintain a stable and comfortable environment for hens.

Housing and space allocation is crucial for supporting the well-being and productivity of laying hens. Overcrowding can lead to stress, increased aggression, and a higher disease incidence, all negatively impacting egg production. Adequate space per bird is essential, with recommended space allowances varying depending on the housing system used. For example, in conventional cage systems, each bird should have at least 67-86 square inches of space, while in cage-free systems, each bird should have a minimum of 1.5 to 2 square feet. Free-range systems require even more space to accommodate natural behaviors such as foraging and dust bathing. Providing perches, nesting boxes, and dust bathing areas also contributes to hens' physical and psychological well-being, encouraging natural behaviors that support overall health and productivity.

Health management is another critical component in enhancing egg production. Diseases, parasites, and other health issues can significantly reduce laying rates and overall flock productivity. Implementing a comprehensive health management program that includes regular health checks, vaccinations, and biosecurity measures is essential. Vaccinations help protect against common poultry diseases such as Newcastle disease, Marek's disease, and infectious bronchitis, which can otherwise lead to severe production losses. Controlling external and internal parasites, like lice, worms, and mites, is essential to maintaining flock health. Parasite burdens can be managed using the proper pesticides and routine deworming.

Stress management is vital for maintaining high levels of egg production. Chickens are sensitive to stress, which can be triggered by various factors such as sudden changes in their environment, handling, loud noises, and social disruption within the flock. Stress can lead to a significant drop in egg production and other health problems. It is essential to maintain a consistent routine, handle birds gently and calmly, and reduce potential environmental stressors to minimize stress. Ensuring that hens can access enrichment activities, such as pecking blocks, foraging opportunities, and dust baths, can also help reduce stress and promote natural behaviors.

Monitoring and record-keeping are essential for identifying and addressing issues affecting egg production. Keeping detailed records of egg production rates, feed consumption, health status, and environmental conditions allows poultry keepers to track trends and identify potential problems early. Regular monitoring of egg production can reveal drops in laying rates that may indicate health issues, nutritional deficiencies, or environmental stressors. Poultry keepers can make informed decisions and take corrective actions to maintain or improve egg production by analyzing these records.

In addition to the basic management practices outlined above, several advanced strategies can further enhance egg production. One such strategy is the use of precision feeding techniques, which involve tailoring the diet to the specific needs of the flock based on factors such as age, laying stage, and health status. Precision feeding can optimize nutrient intake, improve feed efficiency, and support peak egg production.

Another advanced strategy is genetic selection and breeding. While commercial hybrid breeds are often used for their high laying potential, selective breeding within a flock can also enhance production traits. By choosing hens and roosters with desirable characteristics such as high laying rates, good egg quality, and disease resistance, poultry keepers can develop a flock well-suited to their specific production goals and environmental conditions.

Technology in poultry farming has also opened new avenues for enhancing egg production. Automated feeding, watering, egg collection, and environmental control systems can improve efficiency and consistency in management practices. For example, computerized lighting systems can precisely control light exposure to optimize laying cycles, while climate control systems maintain stable temperature and ventilation conditions. Data collection and analysis tools can also provide real-time insights into flock performance, helping poultry keepers make data-driven decisions.

Biosecurity measures are critical in preventing the introduction and spread of diseases within a flock. These measures include controlling access to the poultry area, disinfecting equipment and footwear, and monitoring the health of new birds before introducing them to the flock. Implementing strict biosecurity protocols can protect hens from infectious diseases that could otherwise devastate egg production.

Water quality and availability are also crucial for maintaining high egg production levels. Clean, fresh water should always be available to hens, as dehydration can quickly lead to a drop in egg production. Water quality should be monitored regularly, and issues like contamination or inadequate supply should be addressed promptly. Additional water sources or cooling systems may be necessary in hot weather to ensure hens remain hydrated and comfortable.

Molting is a natural process in chickens, during which they shed and regrow their feathers. Hens typically reduce or stop laying eggs during this period as their bodies divert energy to feather regeneration. While molting is ordinary and necessary, it can impact egg production. Managing molting through controlled lighting and nutrition can help minimize its effects and support a quick return to laying.

Addressing behavioral issues within the flock is important for maintaining harmony and productivity. Feather pecking, aggression, and cannibalism can arise in stressed or overcrowded flocks and significantly impact egg production. Providing adequate space and enrichment and managing flock dynamics through practices such as beak trimming or using pecking deterrents can help mitigate these issues.

Marketing and financial factors are important factors for commercial egg farmers when trying to increase egg production. Understanding market demands, consumer preferences, and economic factors such as feed costs and egg prices can inform production strategies and decisions. Diversifying product offerings, such as producing specialty eggs (e.g., organic, free-range, omega-3 enriched), can also add value and appeal to niche markets, potentially increasing profitability.

Sustainability practices are increasingly important in modern poultry farming. Implementing environmentally friendly practices such as reducing waste, using renewable energy sources, and improving feed efficiency not only benefits the environment but can also enhance the long-term viability of egg production. Sustainable practices can attract environmentally conscious consumers and open up new market opportunities.

Community and educational outreach can also enhance egg production. Engaging with the local community, participating in agricultural education programs, and sharing knowledge and best practices with other poultry keepers can foster a supportive network and contribute to the overall improvement of poultry farming practices. Collaboration and knowledge sharing can lead to innovations and advancements that benefit the entire industry.

In conclusion, enhancing egg production involves a holistic approach encompassing genetics, nutrition, environmental management, health care, stress reduction, monitoring, and advanced strategies. By understanding and optimizing each of these factors, poultry keepers can achieve higher levels of egg production, improve the well-being of their hens, and ensure the sustainability and profitability of their operations. Whether in a commercial setting or a backyard flock, the principles and practices outlined in this section provide a comprehensive guide to achieving success in egg production. As the demand for eggs continues to grow, the importance of efficient and sustainable egg production practices cannot be overstated, making the knowledge and application of these strategies more relevant than ever.

Proper Egg Collection and Storage

As a poultry keeper, farmer, or individual involved in egg production or backyard farming, your role in proper egg collection and storage is crucial. These practices are not just about maintaining egg quality, safety, and freshness, but also about preserving the integrity of the eggs from the moment they are laid until they reach consumers' tables. In this section, we'll explore the importance of your role in implementing proper egg collection and storage techniques, as well as best practices to ensure optimal egg quality throughout the handling process.

The first step in proper egg collection is establishing a simple routine for gathering eggs from the nesting boxes. Eggs should be collected at least once a day, preferably more frequently in hot weather, to prevent spoilage and reduce the risk of contamination. Early morning or late evening is often the best time for egg collection, as it minimizes disturbance to the flock during peak laying hours. When collecting eggs, remember to handle them gently and avoid rough handling or dropping, as this can lead to cracks or breakages. With these simple steps, you can ensure the quality of your eggs.

Nesting boxes should provide a clean, dry, and comfortable environment for hens to lay their eggs. Keeping the nesting boxes free of debris, manure, and other contaminants that could spoil the eggs is essential. Regular cleaning and maintenance of the nesting boxes help ensure that eggs remain clean and free from contamination. Adding clean bedding material, such as straw or wood shavings, can help absorb moisture and provide a comfortable surface for the hens to lay their eggs.

After collecting the eggs, they should be inspected for any abnormalities or signs of damage. Cracked, dirty, or misshapen eggs should be discarded, as they may pose a risk to food safety and quality. Eggs with cracks or breaks in the shell are more susceptible to bacterial contamination and spoilage, so removing them from the batch is crucial to prevent contamination of the remaining eggs. Additionally, misshapen eggs may indicate underlying health issues in the hens, so monitoring for abnormalities can help identify potential problems early on.

Once the eggs have been inspected and sorted, they should be stored promptly in a clean and relaxed environment to maintain their freshness and quality. Ideally, eggs should be stored at around 50-60 degrees Fahrenheit with moderate humidity. Storing eggs at temperatures lower than this can cause them to freeze, while higher temperatures can accelerate spoilage. It is important to avoid storing eggs near strong-smelling foods or substances, as eggs can absorb odors easily.

Egg cartons are commonly used for storing and transporting eggs, as they protect against breakage and contamination. When placing eggs in cartons, care should be taken to position them with the pointed end facing downwards. That helps to keep the air cell, located at the blunt end of the egg, centered and prevents it from shifting during storage. It is also important to avoid overcrowding the eggs in the carton, as this can increase the risk of breakage and spoilage.

Labeling eggs with the collection date can help track their freshness and ensure they are used within a reasonable timeframe. While eggs can remain fresh for several weeks when stored properly, for optimal quality, it is best to use them within three to four weeks of collection. As eggs age, their quality gradually declines, so using them promptly helps to ensure they are at their best when consumed.

In addition to proper handling and storage techniques, washing eggs is a debate among poultry keepers. While washing can remove dirt and bacteria from the eggshell, it also removes the protective bloom. This natural coating helps seal the shell's pores and prevent bacterial contamination. If necessary, it should be done with warm water (about 10 degrees warmer than the egg) and a mild detergent approved for egg washing. The eggs should be rinsed thoroughly and dried completely before storage to prevent moisture from promoting bacterial growth. Alternatively, some poultry keepers prefer to dry clean eggs using a dry brush or sandpaper to remove dirt and debris. Dry cleaning preserves the protective bloom while still effectively removing surface contaminants. Whichever method is chosen, it is essential to handle eggs gently and avoid using abrasive materials that could damage the shell.

In commercial egg production, eggs are often sanitized before packaging to reduce the risk of contamination further. It involves washing the eggs with sanitizer and drying them thoroughly before packaging. Sanitizing eggs helps ensure they meet food safety standards and remain free from harmful bacteria. Proper egg storage is crucial for maintaining egg quality and safety, but handling eggs safely and hygienically during preparation and cooking is equally essential. Eggs should be refrigerated promptly after purchase and stored in their original carton to maintain freshness and prevent contamination. When using eggs in recipes, care should be taken to avoid cross-contamination with other foods, and eggs should be cooked thoroughly to kill harmful bacteria.

In conclusion, proper egg collection and storage are essential practices in poultry management to ensure egg quality, safety, and freshness. By following best practices for handling, inspecting, and storing eggs, poultry keepers can maintain the integrity of their eggs from the nest to the table. Whether for commercial production or backyard farming, attention to detail and proper hygiene are crucial to producing high-quality eggs that meet food safety standards and consumer expectations.

CHAPTER X

Seasonal Care and Management

Adapting to Seasonal Change

Adapting to seasonal changes is a critical aspect of life for chickens, as these birds are highly attuned to environmental cues that signal shifts in weather, daylight, and food availability. From physiological adjustments to behavioral changes, chickens employ a variety of adaptations to thrive in different seasons. To ensure their flock's health, welfare, and productivity, understanding how chickens adapt to seasonal changes is essential for poultry keepers. This section explores how chickens adjust to seasonal changes, from breeding and molting to foraging behavior and housing management changes.

One of the most significant ways chickens adapt to seasonal changes is through changes in their reproductive behavior. Chickens are susceptible to changes in daylight length, a primary cue for regulating their reproductive cycles. As the days grow longer in spring and summer, hens naturally increase their egg production and may even go broody or exhibit a strong desire to incubate eggs and raise chicks. This increase in reproductive activity is driven by hormonal changes triggered by longer daylight hours, which signal to hens that conditions are favorable for breeding and raising young. Conversely, as daylight hours decrease in autumn and winter, egg production typically declines, and hens may enter a period of reduced reproductive activity. Understanding these seasonal fluctuations in egg production allows poultry keepers to anticipate changes in flock management and adjust feeding, lighting, and housing accordingly.

Molting is another vital adaptation to seasonal changes observed in chickens. Molting is the process by which chickens shed and regrow their feathers, typically occurring yearly in response to changes in daylight length and temperature. During molting, chickens may temporarily cease egg production as their bodies redirect energy and resources towards feather regrowth. Molting allows chickens to replace damaged or worn feathers and maintain their insulative properties significantly as temperatures drop in autumn and winter. Poultry keepers can support chickens during molting by providing a balanced diet of protein, vitamins, and minerals to support feather regrowth and overall health. Additionally, ensuring that chickens have access to a clean and comfortable environment can minimize stress and support their recovery during this period of physiological change.

Behavioral adaptations also play a crucial role in how chickens cope with seasonal changes. Chickens are opportunistic foragers, and their foraging behavior can vary depending on seasonal factors such as food availability and weather conditions. In spring and summer, when food sources are abundant and temperatures are mild, chickens may spend more time foraging outdoors, exploring their environment, and engaging in social behaviors like dust bathing and sunbathing. As temperatures drop in autumn and winter, chickens may become less active and spend more time seeking shelter and warmth. Poultry keepers can support chickens during colder months by providing adequate shelter, bedding, and protection from predators, as well as supplementing their diet with nutritious treats to help maintain body condition and energy levels.

Housing management is another critical consideration for poultry keepers adapting to seasonal changes. Chickens require shelter from the elements, particularly during extreme weather conditions such as heatwaves, cold snaps, and storms. In warmer months, providing adequate ventilation and shade helps chickens stay relaxed and comfortable, while in colder months, ensuring insulation and protection from drafts helps chickens stay warm and dry. Poultry keepers may also need to adjust lighting schedules to mimic natural daylight patterns, particularly in winter when daylight hours are shorter. Supplemental lighting can help maintain egg production and prevent disruptions to the flock's reproductive cycle. Additionally, providing access to fresh water year-round is essential for chickens' health and well-being, as dehydration can occur in hot and cold weather.

Chickens also exhibit social adaptations to seasonal changes, particularly in interacting with their flock mates and responding to environmental changes. Chickens are social animals that form hierarchies within their flocks, with dominant individuals asserting their authority over subordinate ones. Seasonal changes can affect social dynamics within the flock, as fluctuations in food availability and environmental conditions may influence competition for resources and social interactions. Poultry keepers should observe their flocks closely during seasonal transitions to ensure that all chickens have access to food, water, and shelter and to monitor for signs of aggression, stress, or illness. Providing enrichment activities such as perches, dust baths, and foraging opportunities can help alleviate boredom and reduce negative social behaviors within the flock.

In conclusion, chickens exhibit a variety of adaptations to cope with seasonal changes, from physiological adjustments to changes in reproductive behavior, molting, foraging behavior, housing management, and social interactions. Understanding these adaptations is essential for poultry keepers to ensure their flock's health, welfare, and productivity year-round. By providing appropriate care, nutrition, and environmental enrichment, poultry keepers can support chickens in adapting to seasonal changes and thrive in diverse environmental conditions. Additionally, monitoring signs of stress, illness, or social conflict allows poultry keepers to intervene promptly and provide the necessary support to maintain a happy, healthy flock.

Winterizing Your Coop and Summer Heat and Management

Winterizing your coop and managing summer heat are essential aspects of poultry care, particularly for chickens, as these birds are sensitive to temperature extremes and require a stable and comfortable environment to thrive year-round. Proper winterization and heat management strategies help poultry keepers ensure their flocks' health, welfare, and productivity during the challenging seasons of winter and summer. This section explores the critical considerations for winterizing your coop to keep chickens warm and comfortable during cold weather and strategies for managing summer heat to prevent heat stress and maintain optimal conditions for poultry.

Winterizing your coop is crucial for protecting chickens from the cold temperatures, drafts, and moisture that can pose health risks and impact egg production during winter. One of the first steps in winterizing your coop is to ensure adequate insulation and ventilation. Insulation helps retain heat inside the coop, while proper ventilation prevents moisture buildup and ammonia levels from rising, which can lead to respiratory issues and frostbite. Insulating the coop's walls, roof, and floor with materials such as foam board, straw, or recycled denim can help maintain a comfortable temperature inside the coop and reduce heating costs. Additionally, installing vents or windows with adjustable covers allows you to regulate airflow and prevent drafts while providing fresh air for the chickens.

Providing adequate bedding is another crucial aspect of winterizing your coop. Deep litter bedding, such as straw, hay, or wood shavings, helps insulate the coop floor and provides a comfortable and dry surface for chickens to roost and nest. Regularly adding fresh bedding and turning it over helps manage moisture and odor, promoting a healthy environment for the flock. Chickens may also benefit from additional heat sources during cold weather, such as heat lamps or heated waterers. However, using and monitoring these heating sources safely is essential to prevent fires or burns.

Managing summer heat is equally essential for ensuring the well-being of chickens and preventing heat-related illnesses such as heat stress and dehydration. Chickens are susceptible to heat stress when temperatures rise above their comfort zone, particularly when combined with high humidity and inadequate ventilation. That helps chickens cope with summer heat; providing adequate shade, ventilation, and access to cool, fresh water is essential. Shade can be provided by natural features such as trees or shrubs or by installing shade cloth or tarps over the coop and run. Ensuring proper ventilation by opening windows, installing fans, or adding vents helps dissipate heat and maintain airflow inside the coop, reducing the risk of heat stress.

In addition to providing shade and ventilation, it's crucial to ensure chickens have access to cool, fresh water at all times during hot weather. Dehydration can occur quickly in chickens exposed to high temperatures, so checking waterers regularly and refilling them with cool water as needed is essential. Adding electrolytes or vitamins to the water can help replenish lost nutrients and support chickens' overall health and hydration levels. Chickens may also benefit from cooling measures such as misters or shallow pans of water for wading or bathing, which help lower their body temperature and relieve the heat.

Managing summer heat also involves adjusting feeding and foraging routines to minimize stress and reduce heat buildup. Feeding chickens during the more excellent parts of the day, such as early morning or late evening, helps reduce metabolic heat production and allows chickens to digest their food more efficiently. Providing shade or shelter near feeding areas encourages chickens to eat and drink regularly, even during the hottest parts of the day. Additionally, offering excellent, refreshing treats such as watermelon, cucumbers, or frozen fruits and vegetables can help hydrate chickens and provide valuable nutrients while offering a welcome reprieve from the heat.

Monitoring chickens closely for signs of heat stress, such as panting, lethargy, drooping wings, or pale combs and wattles, is essential during hot weather. If heat stress is suspected, cooling the chickens and preventing further complications is crucial. Moving chickens to a shaded or ventilated area, providing excellent water for drinking and bathing, and offering electrolytes or ice cubes can help lower their body temperature and alleviate heat stress. In severe cases, chickens may require veterinary attention or supportive care to recover from heat-related illnesses.

In conclusion, winterizing your coop and managing summer heat are essential aspects of poultry care for chickens. Poultry keepers can protect their flocks from cold temperatures and ensure their comfort and well-being by providing adequate insulation, ventilation, bedding, and heating sources during winter. Similarly, by providing shade, ventilation, cool water, and appropriate feeding and foraging opportunities during summer, poultry keepers can prevent heat stress and dehydration and maintain optimal conditions for their chickens. Monitoring the flock closely for signs of discomfort or illness and taking prompt action to address any issues are crucial for ensuring chickens' health, welfare, and productivity year-round. With proper care and management, chickens can thrive in diverse environmental conditions and continue to provide eggs, meat, and companionship to their keepers.

Ensuring Year-Round Coop Comfort and Productivity

Ensuring year-round coop comfort and productivity for chickens is a multifaceted task that requires careful planning, attention to detail, and proactive management strategies. Chickens are resilient and adaptable creatures but rely on their environment for shelter, nutrition, and protection from the elements. Poultry keepers can create a comfortable and productive environment for their flocks throughout the year by implementing a comprehensive approach to coop management.

One of the foundational aspects of ensuring year-round coop comfort and productivity is providing a well-designed and appropriately sized coop. The coop serves as the chickens' primary shelter, offering protection from predators, inclement weather, and extreme temperatures. Ideally, the coop should be spacious enough to accommodate the size of the flock comfortably, with ample room for roosting, nesting, and foraging. Adequate ventilation is essential for maintaining air quality and regulating temperature and humidity levels inside the coop. Windows, vents, and adjustable openings allow airflow, preventing drafts and moisture buildup. Additionally, ensuring the coop is secure and predator-proof helps keep chickens safe and reduces stress, allowing them to focus on laying eggs and thriving.

Another critical factor in coop comfort and productivity is bedding management. Bedding materials such as straw, hay, wood shavings, or shredded paper provide insulation, absorb moisture, and create a comfortable surface for chickens to roost and nest. Regularly cleaning and replacing bedding helps maintain a clean and hygienic environment, reducing the risk of respiratory issues, parasites, and infections. Deep litter systems, where fresh bedding is added regularly to build up a deep layer of composting material, can help regulate temperature and humidity levels inside the coop and provide natural insulation during colder months. Properly managed bedding also reduces odors and ammonia levels, promoting healthier conditions for chickens and improving egg quality.

In addition to coop design and bedding management, nutrition is crucial in ensuring chickens' year-round coop comfort and productivity. A balanced and varied diet is essential for supporting chickens' overall health, immune function, and egg production. Commercially formulated layer feeds provide the vital nutrients, vitamins, and minerals that laying hens need to lay eggs consistently. Supplementing the diet with fresh fruits, vegetables, grains, and protein sources such as mealworms, kitchen scraps, or surplus garden produce adds variety and enrichment to chickens' diets. Additionally, providing access to grit and oyster shells helps chickens digest their food correctly and maintain solid and healthy eggshells.

Lighting management is another critical consideration for year-round coop comfort and productivity. Chickens are sensitive to changes in daylight length, which can influence their reproductive cycles and egg-laying behavior. Providing supplemental lighting in the coop during the shorter days of fall and winter helps maintain consistent egg production and prevent disruptions to the flock's reproductive cycle. Timers can be used to simulate natural daylight patterns, ensuring that chickens receive the appropriate amount of light each day. It's essential to balance providing enough light to stimulate egg production and allowing chickens to rest and maintain their natural circadian rhythms. Monitoring chickens' behavior and egg production helps poultry keepers adjust lighting schedules to meet the flock's needs.

Managing environmental stressors is crucial for ensuring chickens' year-round coop comfort and productivity. Extreme temperatures, inclement weather, and predator threats can all cause stress and impact chickens' health and egg-laying performance. Providing adequate shelter, ventilation, and insulation helps chickens regulate their body temperature and stay comfortable in all weather conditions. Installing predator-proof fencing, locks, and hardware protects chickens from predators such as raccoons, foxes, and birds of prey, allowing them to feel secure and focused on foraging and laying eggs. Additionally, enrichment activities such as perches, dust baths, and foraging opportunities help alleviate boredom and reduce stress, promoting natural behaviors and overall well-being.

Routine health care and disease prevention are essential components of year-round coop management. Regular health checks, vaccinations, and parasite control measures help prevent diseases and ensure chickens remain healthy and productive. Monitoring chickens for signs of illness or injury allows poultry keepers to intervene promptly and provide the necessary care and treatment. Fresh birds should be quarantined before being added to the flock to stop the spread of parasites and diseases. Good hygiene habits help lower the risk of diseases and create a healthy environment for hens. These activities include routinely cleaning and sanitizing the coop, removing feces and dirty bedding, and giving the birds clean water and feed.

In conclusion, ensuring year-round coop comfort and productivity for chickens requires a comprehensive approach that addresses coop design, bedding management, nutrition, lighting, environmental stressors, health care, and disease prevention. By providing a well-designed and appropriately sized coop, managing bedding effectively, offering a balanced diet, regulating lighting, managing environmental stressors, and practicing routine health care, poultry keepers can create a comfortable and productive environment for their flocks throughout the year. With careful planning, attention to detail, and proactive management strategies, poultry keepers can support chickens' health, well-being, and egg-laying performance, leading to a rewarding and successful poultry-keeping experience.

CHAPTER XI

Troubleshooting Common Issues

Addressing Common Problems in Chicken Keeping

Roosters rule the roost, but when they start attacking people, something has to be done. The problem often begins when roosters perceive your actions as aggressive. Wearing floppy boots or swinging a bucket may seem like a challenge in your rooster's eyes, and as the guardian of the flock, he won't back down from a challenge. To tame an attack rooster, catch and hold him whenever he starts acting up. Make sure you're wearing protective clothing. Hold him until he calms down and realizes he isn't going anywhere and you're in charge. Be persistent. This method may work better with young roosters rather than older birds who are more set in their ways.

Chicken owners are often concerned when one or more birds experience feather losses, commonly accompanied by slowed or halted egg production. There's nothing to worry about. According to Kansas State University Extension, molting is most often controlled by the bird's hormones, which are regulated by the amount of light they receive each day. Hens will molt at about one year, then often molt in the fall. If the feather loss is due to molting, you'll be able to see new feathers coming in, but feather loss also occurs for other reasons. Nutrition may be an issue. Mites or lice may irritate. As they grow, broiler chicks often have patches of bare skin since they gain weight faster than the feathers can come in. Hens may also lose feathers on their backs during mating. If feather loss results from pecking, reduce the intensity of light in the coop.

It is an alarming problem for chicken owners. Cannibalistic birds peck at each other's feathers, toes, heads, and bodies. Because birds naturally imitate each other, it can spread quickly through a flock once this behavior starts. If not managed quickly, flesh injuries and death can result in the loss of much of a flock. According to Kansas State University Extension, excessive lighting, overheating, poor nutrition, overcrowding, and intermingling birds that have not been reared together can lead to cannibalism. To control this nasty habit, consider hiring an expert to trim the offending birds' beaks, remove injured and aggressive birds, dim the light, turn birds outside, or provide scratch grain or grass clippings in the pen. You can also try hanging CDs or other shiny objects at different heights to distract the birds and give them something else to peck at.

It's a common problem -- chickens who decrease egg production or stop laying eggs altogether. The experts at Virginia Cooperative Extension say the most common causes are decreasing day length, improper nutrition, disease, advancing age (2 or 3 years), and stress. Hens require 14 hours of light daily to sustain egg production, so provide artificial light as days become shorter. Ensure they have a constant supply of fresh water, layer food with 16% to 18% protein, avoid whole grains and scraps, and offer oyster shells. Contact a veterinarian if you suspect disease, and keep new birds isolated from the rest of the flock. Reduce stress by avoiding moving, handling, fright, and changing environmental conditions.

Once laying hens get a taste of eggs, it's difficult, if not impossible, to break them of the habit. That's why managing your flock is essential, so the hens never get that first taste of delicious eggs! If broken eggs are in the coop, a hen will likely eat them. Virginia Cooperative Extension says to avoid excessive traffic in the nesting area and supply plenty of clean, dry nesting materials. Remove broody hens from the nesting area. Keep eggshells strong by feeding a complete ration and offering oyster shells. Reduce stress by keeping bright lights out of coops, especially near the nesting area. Also, avoid sudden movement in the nesting area and scaring hens out of their boxes. Collect eggs early in the day and often, and if you do have an egg-eater, cull her from the flock before the others follow suit.

You can provide your hens with the most beautiful, comfortable coop imaginable, but they may still need to lay their eggs there. If your birds can go outside, they may find their places to roost and nest. Trees are perfect for roosting, and they may make their nest and lay eggs under a thorny bush or inside a hollow tree. Once a chicken finds a place it likes, getting them to change their minds is very difficult. If they roost in trees, it's essential to at least get them in at night, even if you must climb the tree. Nocturnal predators like owls and raccoons pose a real threat after dark. If your birds decide to roost or lay in places other than their coop and nest box, chances are you will have to live with it.

She naturally tends to sit on eggs to hatch them when a hen lays eggs, even unfertilized ones. When a nest is full of eggs results in hormonal changes that result in the cessation of laying eggs; it has become broody. Some hens will go broody even without this stimulus, however. According to Oregon State University Extension, Cochins and Silkies are known for going broody, while Leghorns rarely do. To reduce broodiness, collect eggs daily. If a hen stays on her nest for several days, remove her and deny access to the nest for several days. Eventually, the broody behavior should stop, and she'll begin laying eggs again.

Parasites like lice and mites can damage a flock, so it's essential to be diligent in detecting and preventing them. Poultry lice can lay up to 300 eggs simultaneously, cemented to the feather shaft. These lice feed on birds' dry skin scales, feathers, and scabs. Infestations occur most often in the fall and winter. Northern Fowl Mites are common, especially in cool climates, and can spread from bird to bird. Chicken mites are common in warmer climates. They suck blood from birds at night and hide in cracks and crevices of the chicken house during the day. Sanitation and cleanliness are vital to controlling lice and mites. Clean and disinfect housing and equipment, reduce traffic through the houses, and avoid contact with wild birds. The Ohio State University Extension says chemical control can include the use of carbaryl. Treat the walls, floors, roosts, nest boxes, and the birds simultaneously, but avoid contact with feed.

Worms are common in backyard flocks, and in large numbers, they can affect growth, egg production, and overall health. According to the University of Florida Extension, chickens pick up parasite eggs by ingesting contaminated feed, water, or litter or eating snails, earthworms, or other insects that carry the eggs. Large roundworms, cecal worms, tiny roundworms, and tapeworms are common. To prevent an infestation, provide a diet rich in vitamins A and B complex, and keep the chickens' area sanitary. Also, avoid overcrowding, use insecticides to control insects, and avoid contact with wild birds. Specific worm infections require specific medical treatments, so see your veterinarian for a diagnosis and the appropriate medication.

Backyard chicken flocks and other poultry are susceptible to coccidiosis, avian influenza, Marek's disease, Newcastle disease, infectious bronchitis, and pox. Several medications can be used to treat diseased chickens, most of which are administered in water or feed. If you suspect your chickens may be diseased, it's essential to contact your veterinarian for recommended treatment.

In conclusion, addressing common problems in chicken keeping requires a proactive and holistic approach that addresses predator attacks, disease outbreaks, parasite infestations, behavioral issues, and flock dynamics. By implementing strict biosecurity measures, practicing good hygiene and sanitation practices, providing adequate space, nutrition, and enrichment, and monitoring chickens' health and behavior closely, poultry keepers can mitigate potential problems and ensure the well-being and productivity of their flocks. With careful management and attention to detail, poultry keepers can enjoy the rewards of raising healthy, happy chickens and reaping the benefits of fresh eggs, vibrant plumage, and a rewarding poultry-keeping experience.

Solutions for Behavioral Issues

Solutions for behavioral issues in chickens are essential for maintaining a harmonious and productive flock. Like any other animals, chickens can exhibit a range of behaviors that may pose challenges for poultry keepers, including feather pecking, cannibalism, egg eating, aggression, and excessive vocalization. Various factors, including overcrowding, boredom, nutritional deficiencies, social stress, or environmental factors, can cause these behaviors. Identifying the underlying causes of behavioral issues and implementing appropriate solutions is crucial for promoting chickens' health, welfare, and well-being.

One common behavioral issue in chickens is feather pecking, a problem that can be effectively addressed with the right solutions. Feather pecking can lead to feather loss, skin damage, and increased susceptibility to injury and disease. To address this problem, you, as a poultry keeper, should provide adequate space, enrichment, and socialization opportunities for chickens to express natural behaviors such as foraging, dust bathing, and roosting. Supplementing chickens' diets with nutritious treats and providing environmental enrichment such as perches, dust baths, and foraging opportunities helps alleviate boredom and reduce the likelihood of feather pecking. Additionally, addressing underlying causes such as overcrowding, poor ventilation, or nutritional imbalances helps prevent feather pecking from occurring in the first place, giving you the confidence that you can effectively manage this issue.

Cannibalism is another common behavioral issue in chickens, where chickens peck and injure each other, often resulting in severe injury or death. Cannibalism can occur for various reasons, including overcrowding, boredom, nutritional deficiencies, or social stress. To address this problem, poultry keepers should provide adequate space, ventilation, and environmental enrichment to reduce stress and promote natural behaviors. Separating aggressive or injured birds from the flock temporarily allows them to heal and prevents further injury. Additionally, distractions such as hanging cabbage or other vegetables, mirrors, or pecking blocks can help redirect chickens' aggressive tendencies and reduce the likelihood of cannibalism.

Egg eating is another common behavioral issue in chickens, where chickens break and consume their or other flock members' eggs. Various factors, including nutritional deficiencies, boredom, insufficient nesting boxes, or accidental egg breakage, can cause egg eating. To address this problem, poultry keepers should ensure that chickens receive a balanced diet with adequate calcium and other nutrients to support strong eggshells. Providing clean and comfortable nesting boxes with plenty of nesting material helps encourage chickens to lay eggs in the designated areas and reduces the likelihood of accidental breakage. Collecting eggs frequently throughout the day prevents them from accumulating in the nest boxes and becoming targets for egg-eating behavior. Additionally, providing distractions such as hanging treats or foraging opportunities helps keep chickens occupied and reduces boredom-related egg-eating behavior.

Aggression is another common behavioral issue in chickens, mainly when introducing new birds or integrating different age groups into the flock. Aggression can lead to injuries, stress, and disruptions in flock dynamics. To address this problem, poultry keepers should introduce new birds gradually, allowing them to acclimate to their surroundings and establish social hierarchies without causing undue stress or conflict. Providing multiple feeding and watering stations helps reduce competition for resources and minimizes aggressive behaviors. Additionally, observing chickens' behavior closely and intervening to separate individuals, if necessary, helps prevent injuries and maintain harmony within the flock. Providing adequate space, ventilation, and environmental enrichment also helps reduce stress and promote positive social interactions among flock members.

Excessive vocalization is another common behavioral issue in chickens, where chickens produce loud and persistent vocalizations that can disrupt poultry keepers and neighbors. Various factors, including boredom, stress, environmental disturbances, or communication with other flock members, can cause excessive vocalization. To address this problem, poultry keepers should provide adequate space, enrichment, and socialization opportunities to keep chickens occupied and stimulated. Providing distractions such as hanging treats, toys, or foraging opportunities helps reduce boredom-related vocalization. Addressing underlying causes such as stress, overcrowding, or environmental disturbances also helps minimize excessive vocalization and promotes a calm and quiet environment for chickens.

Addressing behavioral issues in chickens is not just a challenge, but also an opportunity for a proactive and holistic approach that can lead to positive outcomes. By providing adequate space, ventilation, nutrition, and environmental enrichment, you, as a poultry keeper, can promote positive behaviors and reduce the likelihood of problematic behaviors such as feather pecking, cannibalism, egg eating, aggression, and excessive vocalization. Monitoring chickens' behavior closely and intervening promptly to address issues as they arise helps maintain a harmonious and productive flock and opens the door to a more rewarding poultry-keeping experience. With your careful management and attention to detail, you can enjoy the rewards of raising healthy, happy chickens and reap the benefits of fresh eggs, vibrant plumage, and a thriving flock.

Handling Environmental Challenges

Handling environmental challenges for chickens is a critical aspect of poultry management. Chickens are highly sensitive to their surroundings and can be affected by a wide range of environmental factors. From extreme temperatures and inclement weather to inadequate housing and poor ventilation, ecological challenges can impact chickens' health, welfare, and productivity. By understanding the potential challenges and implementing appropriate solutions, poultry keepers can create a comfortable and safe environment for their flocks.

One of the primary environmental challenges for chickens is temperature regulation, particularly in regions with extreme weather conditions. Chickens are sensitive to heat and cold stress, and inadequate temperature control can lead to heatstroke, hypothermia, and other health issues. To address this challenge, poultry keepers should provide adequate shelter and insulation to protect chickens from extreme temperatures. In hot weather, providing shade, ventilation, and access to cool water helps chickens stay cool and hydrated. Installing fans, misters, or evaporative cooling systems in the coop helps reduce heat stress and maintain comfortable temperatures. In cold weather, providing supplemental heat sources such as heat lamps, radiant heaters, or heated perches helps chickens stay warm and prevent frostbite. Insulating the coop walls and roof, providing ample bedding, and sealing drafts help retain heat and maintain a comfortable temperature inside the coop.

Another environmental challenge for chickens is adequate housing and ventilation. Chickens require well-designed and adequately maintained housing to protect them from predators, inclement weather, and environmental stressors. Inadequate ventilation can lead to poor air quality, moisture buildup, and respiratory issues in chickens. To address this challenge, poultry keepers should ensure the coop is well-ventilated, with windows, vents, and adjustable openings that allow airflow without causing drafts. Properly positioned ventilation openings help remove excess moisture, ammonia, and airborne pathogens from the coop, promoting a healthy environment for chickens. Regularly cleaning and disinfecting the coop, removing soiled bedding, and providing clean water and feeders help maintain hygiene and prevent disease transmission.

Parasite infestations are another environmental challenge that poultry keepers may encounter. External parasites such as mites, lice, and ticks can cause irritation, discomfort, and stress in chickens, leading to decreased egg production and poor feather quality. Internal parasites such as worms can also affect chickens' overall health and productivity. To address this challenge, poultry keepers should implement a comprehensive parasite control program that includes regular inspections, preventative treatments, and targeted interventions as needed. Providing clean and dry bedding, regular dust baths, and access to diatomaceous earth or other natural pest control methods helps reduce the risk of parasite infestations and keeps chickens healthy and comfortable.

Inadequate nutrition is another environmental challenge impacting chickens' health and productivity. Chickens require a balanced diet with adequate protein, vitamins, and minerals to support growth, egg production, and overall health. Inadequate nutrition can lead to poor egg quality, reduced fertility, and increased susceptibility to disease. To address this challenge, poultry keepers should provide chickens with a high-quality commercial feed formulated specifically for their age and stage of production. Supplementing the diet with fresh fruits, vegetables, grains, and protein sources such as mealworms or kitchen scraps adds variety and enrichment to chickens' diets. Additionally, providing access to grit and oyster shells helps chickens digest their food correctly and maintain solid and healthy eggshells.

Noise pollution is another environmental challenge impacting chickens' health and welfare. Loud noises from nearby roads, construction sites, or industrial activities can cause stress and disrupt chickens' natural behaviors, decreasing egg production and increasing susceptibility to disease. To address this challenge, poultry keepers should minimize noise disturbances around the coop and provide a quiet and peaceful environment for chickens to roost, forage, and socialize. Installing sound barriers, planting dense vegetation, or creating natural buffers helps reduce noise pollution and create a more tranquil environment for chickens.

In conclusion, handling environmental challenges for chickens requires a proactive and holistic approach that addresses temperature regulation, housing and ventilation, parasite control, nutrition, and noise pollution. By understanding the potential challenges and implementing appropriate solutions, poultry keepers can create a comfortable and safe environment for their flocks, promoting their health, welfare, and productivity. With careful management and attention to detail, poultry keepers can enjoy the rewards of raising healthy, happy chickens and reap the benefits of fresh eggs, vibrant plumage, and a rewarding poultry-keeping experience.

CONCLUSION

In conclusion, "Raising Chickens for Beginners: From Coop Design to Egg Collecting" provides a comprehensive guide for novice poultry keepers embarking on their chicken-raising journey. Throughout this book, readers are equipped with essential knowledge and practical advice to successfully raise healthy, happy chickens and enjoy the rewards of fresh eggs, vibrant plumage, and a fulfilling poultry-keeping experience.

From the initial steps of planning and designing a suitable coop to the intricacies of chicken care, nutrition, and health management, this book has covered every aspect of raising chickens with clarity and detail. Readers have learned how to select the right breed for their needs, create a comfortable and secure living environment, and provide proper nutrition and healthcare for their flock. Additionally, valuable insights into understanding chicken behavior, managing social dynamics, and troubleshooting common problems have empowered readers to address challenges as they arise confidently.

Moreover, this book has emphasized the importance of ethical and sustainable practices in chicken keeping, promoting the well-being and welfare of chickens as sentient beings. By fostering a deeper understanding of chickens' natural behaviors and needs, readers have been encouraged to cultivate a respectful and compassionate relationship with their flock, enriching their lives and their feathered companions' lives.

As readers embark on their chicken-raising journey with the knowledge and skills gained from this book, they are poised to experience the joys and rewards of nurturing a flock of chickens from hatchlings to productive layers.

Thank you for buying and reading/ listening to our book. If you found this book useful/ helpful please take a few minutes and leave a review on the platform where you purchased our book. Your feedback matters greatly to us.